THE MASTER ARCHITECT SERIES
MURPHY/JAHN
Selected and Current Works

世界建筑大师优秀作品集锦
墨菲／扬建筑师事务所

蒋家龙　刘俊　詹晓薇　译

中国建筑工业出版社

著作权合同登记图字：01-2003-0645号

图书在版编目（CIP）数据

墨菲/扬建筑师事务所/澳大利亚 Images 出版集团编；蒋家龙等译．
—北京：中国建筑工业出版社，2004
（世界建筑大师优秀作品集锦）
ISBN 7-112-06694-8

Ⅰ．墨... Ⅱ．①澳...②蒋... Ⅲ．建筑设计－作品集－澳大利亚－现代 Ⅳ．TU206

中国版本图书馆 CIP 数据核字（2004）第 058090 号

Copyright © The Images Publishing Group Pty Ltd
All rights reserved. Apart from any fair dealing for the purposes of private study, research, criticism or review as permitted under the Copyright Act, no part of this publication may be reproduced, stored in a retrieval system or transmitted in any form by any means, electronic, mechanical, photocopying, recording or otherwise, without the written permission of the publisher. and the Chinese version of the books are solely distributed by China Architecture & Building Press.

本套图书由澳大利亚 Images 出版集团有限公司授权翻译出版

责任编辑：程素荣
责任设计：郑秋菊
责任校对：赵明霞

世界建筑大师优秀作品集锦
墨菲/扬建筑师事务所
蒋家龙　刘　俊　詹晓薇　译

中国建筑工业出版社出版、发行（北京西郊百万庄）
新　华　书　店　经　销
北京嘉泰利德公司制版
恒美印务有限公司印刷
＊
开本：787×1092毫米　1/10　印张：26⅖　字数：660千字
2005年1月第一版　　2005年1月第一次印刷
定价：**218.00**元
ISBN 7-112-06694-8
　TU·5848（12648）
版权所有　翻印必究
如有印装质量问题，可寄本社退换
（邮政编码 100037）
本社网址：http://www.china-abp.com.cn
网上书店：http://www.china-building.com.cn

Contents

目 录

7 导言
　　为城市边缘而设计
　　罗斯·米勒

作品精选
13　网架式建筑
47　塔式建筑
131　城市街区建筑
173　交通建筑

事务所简介
222　墨菲/扬建筑师事务所
223　穹顶屋
224　传记
225　共事人及合作者
228　建筑及项目年表
243　获奖项目及展览
253　参考文献
261　致谢
262　索引

Introduction

Designing for the Urban Edge
By Ross Miller

导　　言

为城市边缘而设计

罗斯·米勒

赫尔穆特·扬的建筑作品第一次受到人们的广泛关注是在 1970 年，当时，他协助解决了在旧建筑的基础之上修建一个新的麦考密克广场[1] 这一充满挑战性的技术难题。在以他的名字命名（墨菲/扬）的芝加哥事务所协助吉恩·萨默尔斯工作的时候，扬就表现出了一种难得的品质，直到今天，这种品质仍体现在他的作品之中：一种从解决问题（即建筑学中的技术方面）到设计本身的完美转换的能力。

该事务所新设了一个计算机室[2]，用以补充原有的两间制图室[3/4]。该事务所更多地依靠模型和书面文件来进行交流，因为扬常常不在办公室，而是在外探讨新的项目或监理在建项目。然而，墨菲/扬事务所的运转情况更像是一个工作室，而非一个具有法人地位的建筑设计公司。建筑设计方案都产生于特定的项目要求，从不先入为主。赫尔穆特·扬也亲自参与到办公室工作的每个阶段之中去。

自 1987 年联合航空公司航站楼[5/6]在芝加哥的奥黑尔国际机场竣工以来，扬一直都在从事机场设计的工作。而且机场项目是了解扬最新设计的一个非常有用的出发点。通过对机场在持续分散化的全球经济中不断演变的功能的分析，扬已经发展出了一些建筑形式，如今，它们正在慕尼黑和科隆被予以精心地阐释。赫尔穆特·扬认识到，机场除了有移动人和货物的作用外，它们已经变成了自给自足的经济和文化节点，即微型城市，而不再是依赖于旧的市中心的卫星城。

在科隆/波恩和慕尼黑机场中心[7]，扬的建筑

Helmut Jahn's architecture first received wide attention in 1970 when he helped solve the challenging technical problem of building a new McCormick Place[1] on the foundations of the old. Assisting Gene Summers at the Chicago firm that now bears his name (Murphy/Jahn), Jahn demonstrated a rare quality that still marks his work today: the ability to move almost seamlessly from problem-solving—the technical side of architecture—to design.

A computer facility[2] has been added to the firm's two existing drafting rooms;[3/4] there is more model making and a greater reliance on written documentation because Jahn is often out of the office presenting new work or supervising ongoing projects; however, Murphy/Jahn still functions more as an atelier than a corporate architectural office. Architectural solutions are generated from the specific demands of the project and never imposed. Helmut Jahn is personally involved with every stage of the work in the office.

Since 1987, when the United Airlines Terminal[5/6] was completed at Chicago's O'Hare International Airport, Jahn has been closely associated with airport design. And airports are a useful point of departure for understanding Jahn's recent work. In analyzing the airport's evolving function in an increasingly decentralized global economy, Jahn has developed certain architectural forms that are now being elaborated upon at Munich and Cologne. Helmut Jahn recognizes that airports, in addition to their role of moving people and freight, have become self-sufficient economic and cultural nodes—micro-cities that are no longer dependent satellites of the old downtowns.

Introduction Continued

思想在一个复杂的微气候受调控的环境之中再现了一个包括了街道、凉廊和建筑正立面等要素的近于完整的都市氛围。这个现代的机场将商业街（现已成为世界性的都市和郊区现象）与传统的运输中心[8]结合在了一起。

举例来说，在慕尼黑机场，连同附近的凯宾斯基大酒店[9]一起，扬已经将他的注意力转移到精炼这座新的"机场城市"的总体城市规划上去了。（参见阿卜杜勒阿齐兹国王国际机场方案和曼谷国际机场方案）。在所谓的"中立地带"中的凯宾斯基大酒店是慕尼黑机场总体规划中的第一个与飞行无关的建筑物。该酒店形成了在新与旧之间的一个充满力量感的空间联系。凯宾斯基大酒店和机场中心都设有巨大的中央房间，打破了在室内和室外空间之间的常规界限。一个几何式的法国风格的花园[10]使得从高技术几何形式到自然形态的转换变得更为从容。在扬的大多数最新的作品中，景观融入到建筑之中，从而产生出一个完全特定的环境。机场从不与自然的世界相对抗，而是努力使其相益得彰。

在科隆/波恩机场中心[11]，需要来回奔波的到港流程通过设置的40个"流通式"柜台和在出发大厅、中央大厅及登机口之间形成的连续性而成为开放式的和无所阻碍的。因此，扬的建筑思想同他发展一座新城市的观念一起，使机场流线简化而更有效率，并将其运输功能变得现代化。这一作品既非先锋派的也不是概念性的，但是它在形式和功能上被高度地精炼化了。

这位建筑师对那些位于主要目的之间的建筑物的密切关注也值得我们注意。这些建筑既非在城镇中，也不在机场，而是位于其边缘。位于巴黎郊外A1高速公路旁的戴高乐机场凯悦大酒店[12]通过运用一些在其他更加复杂的机场方案设计中所磨练出来的那种内向性的技巧来构成其本身的尺度。一个巨大的花园中庭将两个5层楼高的酒店建筑物分隔开来，同时也在一个被放大了的尺度上保留了巴黎式的庭院旅馆所具有的某些亲密感。这种自我参照式的城市主义被应用于斯图加特的帕拉斯办公大楼[13]和慕尼黑的订购中心[14]。

7

8

9

10

11

12

At Cologne/Bonn and Munich Airport Center[7], Jahn's architecture recreates a nearly complete urban ambience of streets, loggias, and building facades within a sophisticated climate-controlled environment. The contemporary airport combines the mall (now a universal urban and suburban phenomenon) with the traditional transportation hub[8].

For example, at Munich, along with the nearby Hotel Kempinski[9], Jahn has turned his attention to refining the overall urban planning of this new "airport city". (See also the King Abdulaziz International and Bangkok International schemes.) Hotel Kempinski in the so-called "Neutral Zone" is the first non-flight-related building in the Munich master plan. The hotel creates a powerful spatial link between new and old. Both Kempinski and the Airport Center have large central rooms that break down normal divisions between interior and exterior space. A formal, French-style garden[10] further eases the transition from high-technology geometry to nature. In most of Jahn's current work, landscape merges into architecture to create a completely specific environment. The airport is never opposed to the natural world but strives to complement it.

At Cologne/Bonn[11] the back-and-forth process of arrival is open and unimpeded with 40 "flow-thru" counters, and a succession of departure halls, concourses, and airplane gates. So Jahn's architecture, as well as developing a new city, also continues to streamline the airport's circulation and modernize its transportation function. The work is neither avant-garde nor conceptual, but highly refined in form and function.

Notable, too, is the architect's close attention to buildings that find themselves between main destinations, neither in town nor at the airport, but at the edge. The Hyatt Regency Roissy[12], located outside of Paris off the A1 Expressway, creates its own scale, using some of the tricks of introversion honed in the more complex airport schemes. The large garden atrium that separates the two five-story hotel blocks preserves some of the intimacy of the

慕尼黑订购中心并不试图向这个位于在城市之外的一个工业区内的鲜为人知的基地作公开的挑战,而是在相联单位的入口庭院之内形成其自己的类似街道景观的外表。交易的"内容"都在"街道"的上面进行,在第二和第三层上发生。由粗大的电缆和张力板所构成的暴露结构系统形成了建筑物在视觉上的多样性。

在提交出他所能做到的最佳工作范例的同时,扬始终挑战着各种建筑类型的极限范围。与这个"边缘"作品同样重要的是,扬在这个时期所作出的最大贡献是在他所获得的那些在城市中的委托项目上,尤其是在柏林的项目。作为他在柏林的第一栋建筑物,库尔菲斯腾达姆 70 号[15]恰如其分地坐落在一条在战后该城市西半部分仅有的遗存了部分真实特征的大街上。所有伟大的建筑——博物馆、歌剧院、市政大楼——都在城市的东部孤零零地存在于废墟之中。库-达姆是一条残存下来的街道,以其夜生活和其与柏林城严格的棋盘格子之间的偏差而著名。扬谨慎地将这个重要的委托看成是一项"城市修复"项目,而并非仅仅是原有的建筑物。

虽然库-达姆 70 号在尺度上大大地缩小了,但它同样需要那些在麦考密克广场设计上所体现出来的罕有的才能。在 20 世纪 50 年代里,在一种被误导了的希望改善交通和步行人流的尝试中,城市规划师们已经将原建筑正立面削掉了 3m。通过一项设计竞赛,扬被留下来完成重建工作。通过赢得特别许可,得以在人行道上悬挑 5m 和超过该城市的 22m 的高度限制,这位建筑师找到了建造他在柏林的第一座塔楼的方法。该塔楼在尺度上显得适宜,使人想起弗兰克·劳埃德·赖特的普赖斯塔楼在细部上的处理,也使人想起门德尔松式的形式。该建筑是这个城市在眼下正经历的重建时期中寻找一个有效的现代风格的第一座建筑物。完成后的建筑物上安装了一个桅杆和冠顶,成为今后的建造者在塑造一种适合于柏林那充满暴力的过去和正在进行着的新生过程的建筑风格时必须面对的一个挑战。虽然扬本人并不关心政治,但他已经无法避免地通过他在柏林的作品而被卷入其中了。

Parisian courtyard hotel at an enlarged scale. This self-referential urbanism is applied to the Pallas Office Building[13] in Stuttgart and the Munich Order Center[14]. The Munich building does not attempt to overtly challenge the anonymity of its site in an industrial park outside the city, but rather creates its own semblance of a streetscape within the entrance courts of the linked units. The "guts" of the business all takes place above the "street", on the second and third floors. Visual variety is created through the exposed structural system of thick cables and tension plates.

Jahn consistently challenges the limits of a type while providing the best working example he can. As important as this "edge" work has been, Jahn's greatest contribution in this period has been in his urban commissions, particularly in Berlin. Kurfuerstendamm 70[15], his first building in Berlin, was appropriately on the only street of any real character left in the western half of the city after the war. All the great architecture—museums, opera houses, civic buildings—were marooned in the East. The Ku-Damm was a leftover street, notable for its nightlife and its deviation from the strict right-angled Berlin grid. Jahn modestly thinks of this important commission less as an original building than as "urban repair."

At a radically reduced scale, Ku-Damm 70 required the same rare mixture of talents as McCormick Place. In a misguided attempt in the 1950s to improve traffic and pedestrian flow, city planners had sheared three meters off the facade of the original building. After a competition, Jahn was retained to complete the restoration. By securing a variance to cantilever five meters over the sidewalk and exceed the city's 22-meter height limit, the architect found a way to build his first Berlin tower. Modest in scale—reminiscent in detailing of Frank Lloyd Wright's Price Tower and also of Mendelsohnian forms—this was the first architecture of the city's current period of rebuilding to find a useful modern idiom. The completed structure, fitted with a mast and crown, is a challenge to future builders to shape an architecture appropriate to Berlin's violent past and ongoing process of rebirth. Although

Introduction Continued

与政治和历史的这种牵连对他的建筑风格来说是件好事。

在没有相当的阻力或设计难度来推敲设计方案的时候，扬所设计的一些商业建筑有时就会在细部上缺乏足够的精致感。柏林自相矛盾的地方在于新的中心是旧的边缘。扬如今已经有能力将在那些在慕尼黑和巴黎郊外的项目上实践而日趋成熟的微型城市主义的许多策略引入到德国从前和未来的首都的建设中来。在斯特拉劳尔广场35号大楼[16]的设计中，他试图让人们回想起那堵将该地块分隔开来达30年之久的混凝土墙，并将它转化成为该项目中一个有序的元素。他用简洁有效的手法将一面墙壁转变成为一个筛网，将该商业建筑群（包括一个重建的地标建筑）所形成的综合功能体变成一系列典型的城市元素——通道、凉廊、门道、玻璃花房、平台、公园和散步道——它们重新将这座城市带到了斯比里河边上。在维多利亚－柏林改造工程[17]中，扬将同样的那种独立而精致的城市主义——这是他在机场项目和斯特拉劳尔广场项目的过程中学到的——应用到了库－达姆上。维多利亚项目是一项精心的修复工程，该项目在一栋现存的外表冷峻的建筑上融合使用了由钢和玻璃构成的一种充满活力的表现手法。扬所设计的这个项目使人想起门德尔松设计的鲜为人知的哥伦布住宅上所体现出来的那种精神，该建筑是在战前完成的，在几十年中它一直都被人们所遗忘。

在20世纪60年代里，柏林市开始尝试恢复那种曾经形成其建筑特征的多样化的文化。辛克尔和门德尔松设计的作品和其他的建筑物一起都在爆炸声中灰飞烟灭了。而那些保留下来的都残存在废墟之中或在前苏联人的管理下被极大地忽略了。文化广场离已经被推平的波茨坦广场这个原来的城市中心和冷战的边缘不远，它被构思成一个用建筑来挑战在东方的文化政治化的现象。汉斯·夏隆设计的柏林爱乐音乐厅和密斯·凡·德·罗设计的国家美术馆表现出两种截然不同的建筑风格，一个是表现派的，显得内向含蓄，而另

personally apolitical, Jahn has inevitably become involved through his Berlin work. This involvement with politics and history has been good for his architecture.

Jahn's commercial buildings have sometimes lacked sufficient refinement of detail when there was not enough critical resistance or technical difficulty to temper the project. The paradox of Berlin is that the new center is the old edge. Jahn has been able to import many of the strategies of micro-urbanism perfected in the outlands of Munich and Paris to the former and future capital of Germany. At Stralauer Platz 35[16], he has managed to recall the concrete wall that bisected the site for 30 years and transform it into an ordering element of the project. He has effectively turned a wall into a sieve, making the mixed-use collection of commercial buildings (including a restored landmark) a catalog of classic urban elements—passage, loggia, gate, winter garden, terrace, park, and promenade—that re-introduce the city to the River Spree. At the Victoria-Berlin[17] renovation, Jahn applies the same sort of self-contained, refined urbanism—learned in the airport schemes and in process at Stralauer—to the Ku-Damm. Victoria is an elaborate repair that marries an exuberant expression of steel and glass to a dour existing building. Jahn's project suggests the spirit of Mendelsohn's lost Columbus Haus, completed before the war and purged, for decades, from civic memory.

In the 1960s the city of Berlin first tried to reach back to the diverse culture that had formed its architectural identity. Buildings by Schinkel and Mendelsohn, among others, had been blasted away. Those that remained were in ruin or horribly neglected under Soviet stewardship. The Kulturforum, off the already leveled Potsdamer Platz (the old center and Cold War edge), was conceived as an architectural challenge to the politicizing of culture in the East. Hans Scharoun's Philharmonie and Mies van der Rohe's National Gallery of Art were two radically different interpretations of architecture. One was expressionist and introverted, the other universal and polemical in its purity. Yet, taken together they reaffirmed the city's faith in itself and the value of imagination over ideology.

一个建筑则是普遍性的,其纯洁性引起广泛的争议。然而,将它们放在一起来看,它们再一次表现出了这座城市对其自身的信心和超出意识形态之外的想像力的价值。

从夏隆的杰作那里就可以看到赫尔穆特·扬设计的面积有2200万平方英尺的索尼中心[18],而后者也从前者那里汲取了很多灵感。一个巨大的漂浮在空中的帐篷顶覆盖着一个"重要空间",该空间大得足以算作室外部分,但却被包围着[19]。索尼中心的一层设有电影院、商业空间以及电子产品展示区,上面是住宅和办公室,使其成为新时期娱乐业的新的波茨坦广场。该建筑计划在1999年12月31日的上午11点对外开放。扬在让一栋建筑适应这种充满刺激性的当代对私密和公共空间[20]的混淆的时候持一种非常认真的态度。索尼中心是一个用于千禧年纪念的文化广场,在这里,严肃的娱乐事业被描绘成对古典音乐和绘画这些纯艺术的一种真正的挑战。

如果不了解扬的作品的另一面就无法看到他的最新设计工作的全貌。除了在柏林项目上所表现出来的便利性和勃勃生机,我们还应看到那些从辛辛那提的喷泉广场西侧项目[21]到在新加坡[22/23]和吉隆坡的那些高层建筑。但直到看到在上海的21世纪塔楼[24]后,我们才能够感受到赫尔穆特·扬正朝着什么方向在努力。这栋在上海的摩天楼不仅有着他早期作品上所具备的便利性,还表现出压倒一切的简洁性。像在外层的玻璃幕墙等这些通常情况下所必需的元素都被小心翼翼地去掉了,直到结构成为建筑物本身:即使它的骨架,也是装饰。景观元素——给人一种不仅是一个格子架,而是热带花园的印象——丰富了这一建筑作品,它们不是简单地为了效果才被附加上去的。作为赫尔穆特·扬最新作品中的佼佼者,其严谨的简洁性正反映着将他的作品所具有的非凡的复杂性。

罗斯·米勒的最新著作是《美国的启示:芝加哥的大火和神话》。在1995年秋,艾佛烈·A·克诺夫将出版他的《城市游戏》,这是一本全面研究都市更新运动之后的美国城市的著作。他的文章曾发表在《华尔街日报》、《进步建筑》、以及《洛杉矶时报》等全国性的出版物上。

22

23

24

Helmut Jahn's 2.2 million square-foot Sony Center[18], within view of Scharoun's masterpiece, takes a good deal of inspiration from it. A huge, floating tent roof covers a "great space"—large enough to be outside but contained[19]. With theaters, businesses, electronic displays on the ground floor, and residences and offices above, the Sony Center is a reclaimed Potsdamer Platz for the new age of entertainment. The building is scheduled to open at 11:00p.m. on December 31, 1999. Jahn is serious about tailoring an architecture to the stimulating contemporary confusion of private and public space[20]. Sony Center is a kulturforum for the millennium in which the serious business of entertainment is portrayed as the real challenge to the high art of classical music and painting.

A portrait of Jahn's recent work would not be complete without an understanding of another side of his work. Along with the facility and exuberance of much of the Berlin work there are the towers, from the Fountain Square West project[21] in Cincinnati, to Singapore[22/23] and Kuala Lumpur. But it is with the 21 Century Tower[24] in Shanghai that one can get the best sense of where Helmut Jahn is heading. The Shanghai skyscraper has all the facility of his earlier work but with a commanding simplicity. Normally necessary elements, such as a wrapping glass wall, are scrupulously removed, until the structure is the building: its skeleton and ornament. Landscape elements—the impression of a tropical garden overtaking a lattice—enrich the architecture; they are not simply added for effect. As in the best of the latest work, disciplined simplicity makes Helmut Jahn's work brilliantly complex.

Ross Miller's most recent book is *American Apocalypse: The Great Fire and the Myth of Chicago*. In Fall 1995 Alfred A. Knopf will publish *City Games*, a comprehensive study of the American city after urban renewal. His writing has appeared in the *Wall Street Journal*, *Progressive Architecture* and the *Los Angeles Times*, among other national publications.

作品精选 / Selected and Current Works

网架式建筑 / Mat Buildings

14	慕尼黑订购中心	Munich Order Center
20	庆典中心	Celebration Center
22	海军码头	Navy Pier
23	圣迭戈会议中心	San Diego Convention Center
24	威斯康星住宅	Wisconsin Residence
28	阿尔贡国立实验室/国家能源部项目辅助研究室	ANL/DOE Program Support Facility
30	二号区警察局总部	Area 2 Police Headquarters
32	德拉加扎职业中心	De La Garza Career Center
34	拉斯特—欧勒姆公司总部	Rust-Oleum Corporation Headquarters
36	圣玛丽体育运动馆	Saint Mary's Athletic Facility
38	密歇根城市公共图书馆	Michigan City Public Library
40	奥拉利亚图书馆	Auraria Library
42	阿布扎比会议中心城	Abu Dhabi Conference Center City
43	明尼苏达州政府和历史中心	Minnesota Government & History Center
44	肯珀竞技场	Kemper Arena

Munich Order Center

Design/Completion 1989/1993
Munich, Germany
Archimedes Gewerbe und Buero Centrum GmbH & Co.
1,500,000 square feet
Composite steel and concrete frame, cable-stayed steel and glass roof
Glass and aluminum curtain wall

慕尼黑订购中心

设计/竣工　　1989/1993
德国，慕尼黑市
阿基米德手工业与办公中心股份有限公司
1,500,000平方英尺
钢与混凝土混合框架结构，由拉索固定的钢和玻璃屋面
玻璃与铝材幕墙

The Munich Order Center (MOC) evolved from a program combining large public exhibit halls for industry shows with small private showrooms (order offices). These two uses have opposite architectural requirements, yet the MOC binds both together with an efficient regular structure and generous public spaces.

The backbone of the building is the service wall in which the mechanical systems, elevators and exit stairs are concentrated. Extruded from the service wall are cable-stayed roofs spanning 78 feet, split at their peaks by skylights. The longest roofs define "solid fingers" which house the order offices on the building's upper two floors and extend the full length of the site. The intervening "open fingers" enclose the building's major spaces, namely the entrance courts, ground-level foyer, first-floor linear atria, and the tree-lined promenade. A bridge spans the promenade, linking the north and south tracts of the building mass.

慕尼黑订购中心（MOC）是一个将用于工业展览的大型公共展厅和小型私人陈列室（订购办公室）结合在一起的计划的产物。这两种功能在建筑上有着截然相反的要求，然而慕尼黑订购中心用一个高效率的规整的结构和宽敞的公共空间将这两者结合在一起。

该建筑的关键部分是一堵设备墙，机械系统、电梯和疏散楼梯都集中布置在这一墙体之中。跨度为78英尺的由拉索固定的屋面从设备墙上延伸出来，在最高处用采光天窗将屋面分隔开。最长的几个屋面勾画出"实体手指"的形状，它们覆盖着位于建筑的上面两层的订购办公室，并且一直延伸贯穿了整个建筑场地。在其之间的"开放手指"围合了该建筑的一些主要空间，包括入口庭院、底层大厅、二层上的线性中庭、以及树木成行的散步道。一座天桥横跨在散步道上，将建筑的南北两个体量联系起来。

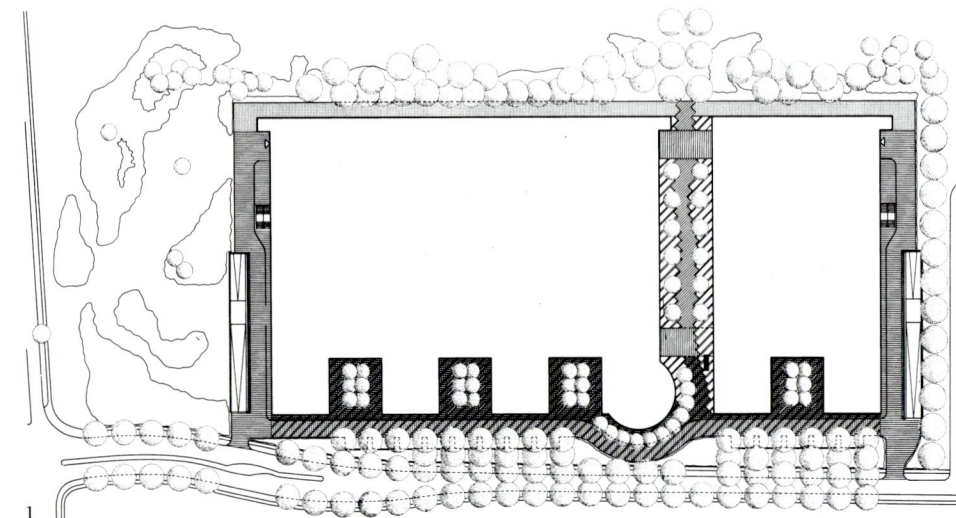

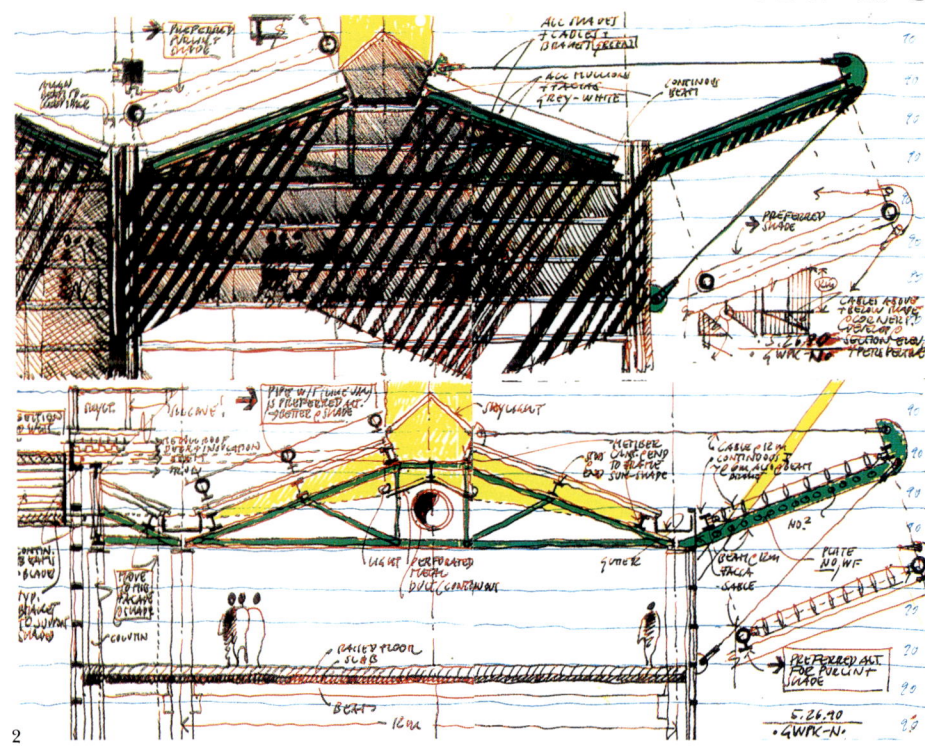

1　Site plan
2　Detail sketch
3　Front facade
4　Restaurant pavilion
5　Entry court
6–7　Sun shades

1　总平面图
2　细部草图
3　正立面
4　餐厅部分
5　入口庭院
6–7　遮阳板

3

4

5

6

7

Munich Order Center

8 Ground-floor plan/sections
9 Louvered service wall
10 Promenade with bridge linking north and south buildings
11 Fritted pattern wall detail

8 底层平面图/剖面图
9 装有百叶的设备墙
10 散步道，上方是连接南北建筑部分的天桥
11 热熔压花玻璃形成的墙面图案细部

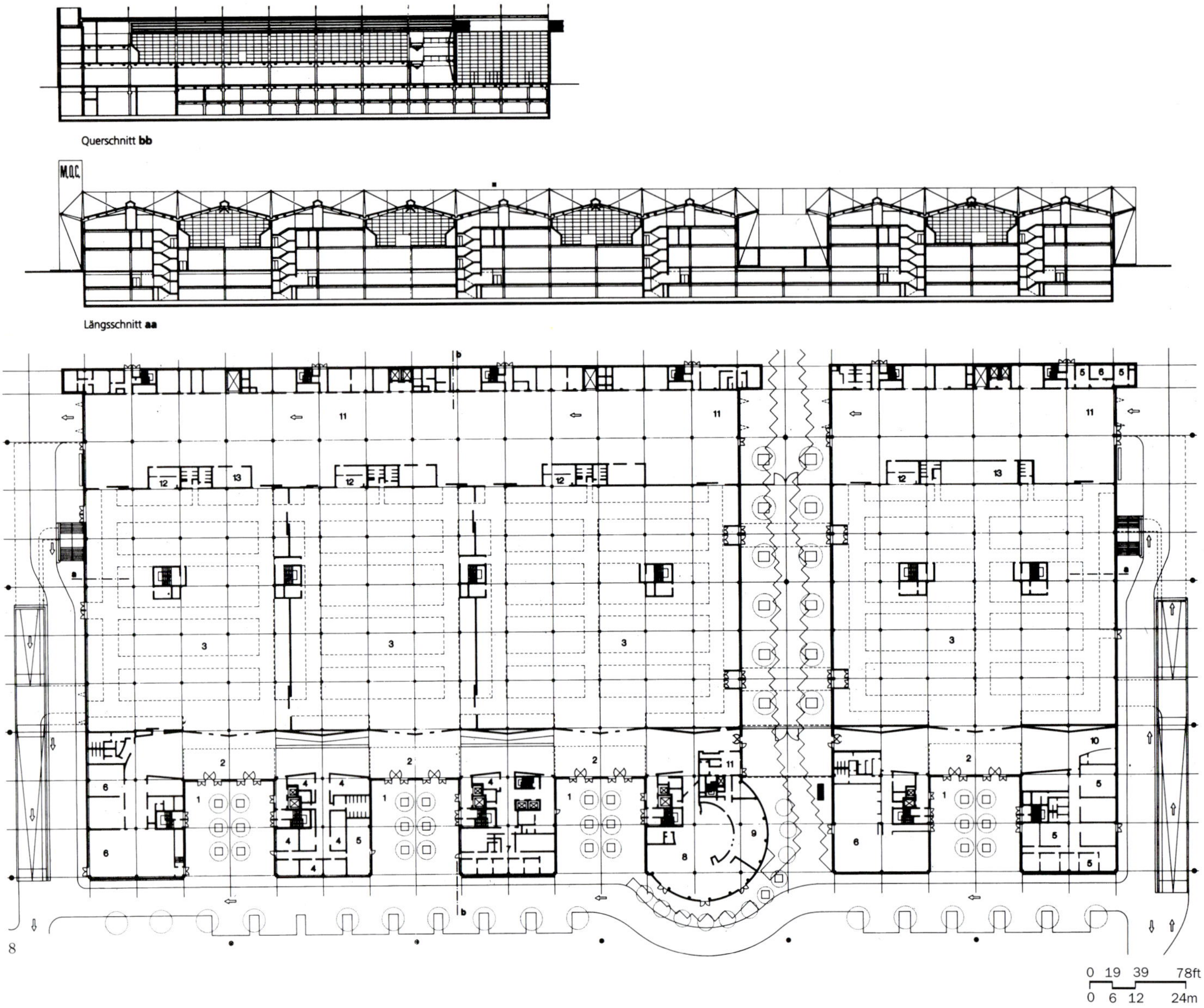

Querschnitt bb
M.Q.C.
Längsschnitt aa
0 19 39 78ft
0 6 12 24m

8

9

10

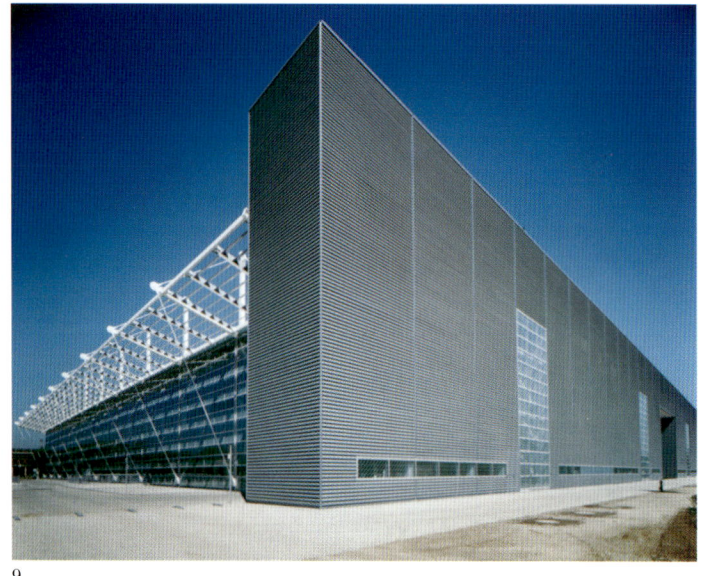

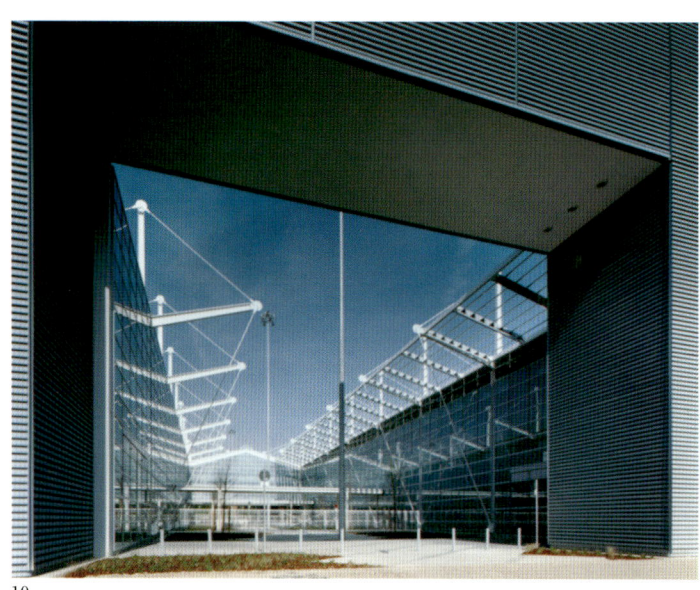

12 a–d Structural/wall details
13 Mast/outrigger/cable tie-down
14 Bridge king post
15 Entrance from second floor with bridge
16 Exhibitors' display area
17 Exhibitors' common area with skylights
18 Seating areas with patterned glass wall
19 Exhibit hall entrance
20 Private office corridor

12 a~d 结构/墙体大样
13 桅杆、悬臂梁、缆索之间的连接固定
14 天桥的主要支撑构件节点
15 二层以天桥作为入口
16 展品陈列区
17 带天窗的展区公共空间
18 带有图案的玻璃幕墙边上的座位区
19 展览大厅入口
20 私人办公室走廊

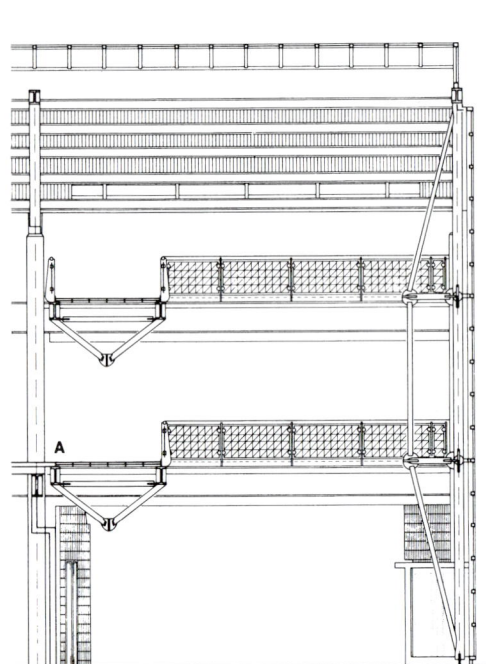

12a

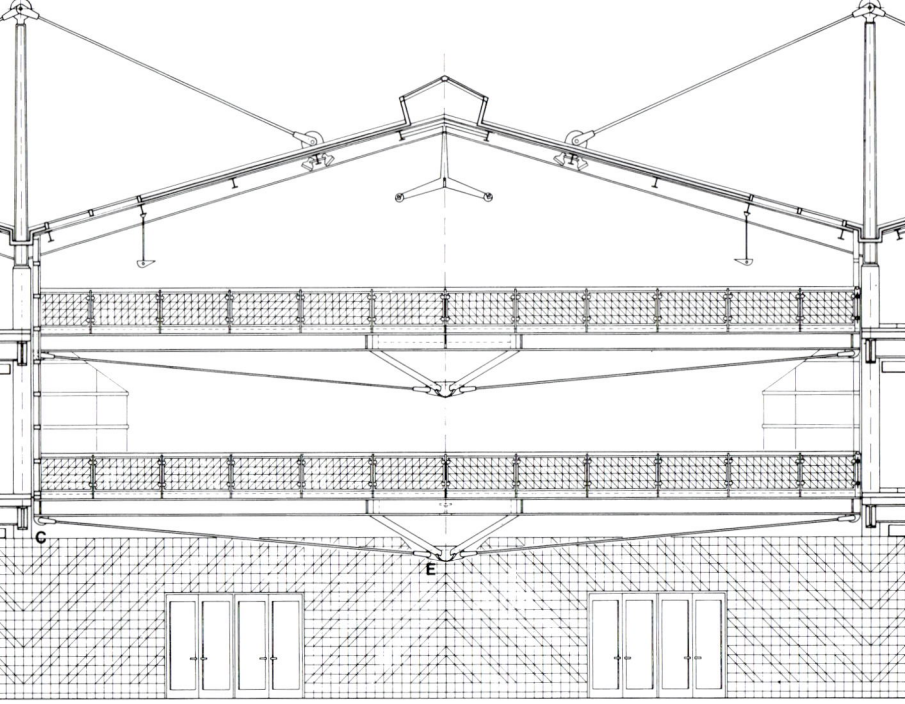

12b

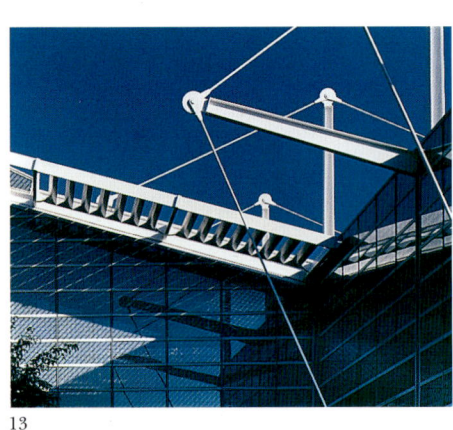

13

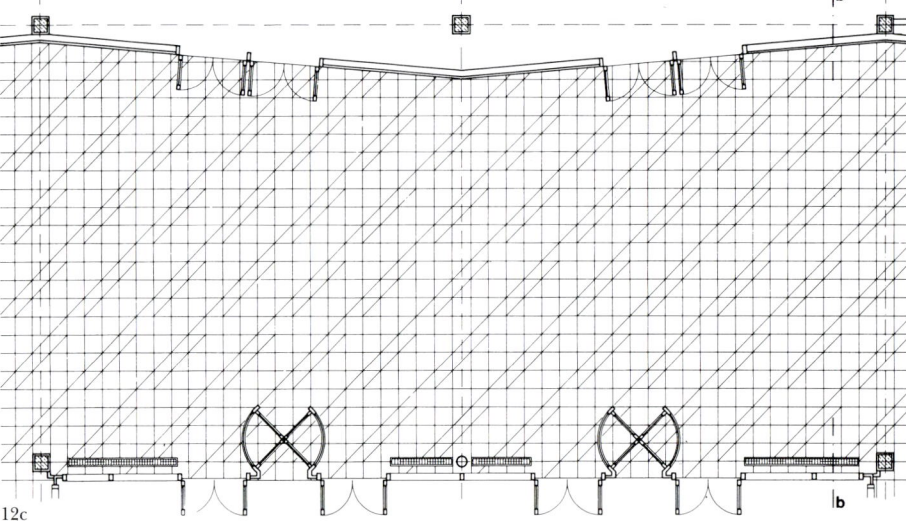

12c

14

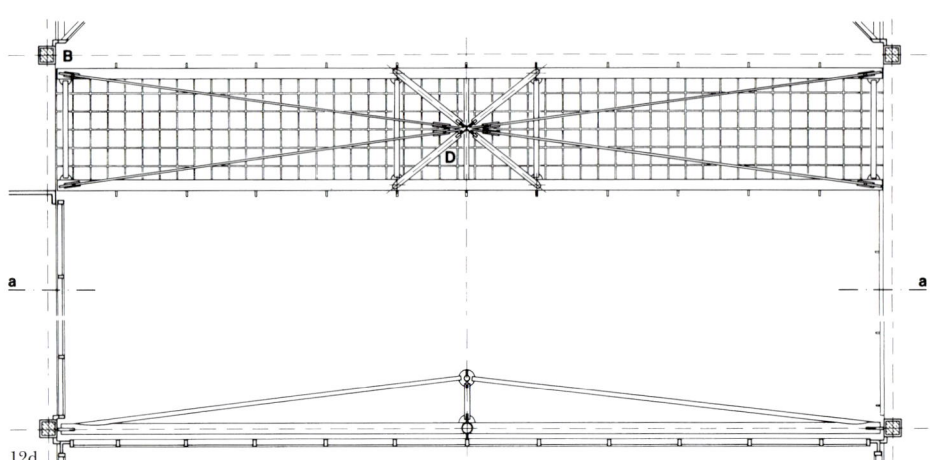

12d

15

16

17

18

19

20

Munich Order Center 19

Celebration Center

Design 1991
Orlando, Florida
Disney Development Co.
1,500,000 square feet
Steel frame
Glass infill, fabric and glass roof

庆典中心

设计　　1991
佛罗里达州，奥兰多市
迪斯尼开发公司
1,500,000平方英尺
钢框架结构
玻璃填充墙，织物和玻璃屋面

Celebration Center is an attempt to rethink the regional mall on functional, technical, aesthetic and experiential levels; to transfer the urban experience of shopping to the suburban realm. Regional shopping centers are generally diagrams of circulation and space management. In contrast, urban shopping is a rich and varied mixture of commerce, culture, recreation and ceremony.

The overall image is of a city surrounded by "green walls" of vines growing on constructed frames, with giant "gates" giving appropriate scale to the various shopping precincts. Beyond the "gates" are the parking garages or surface parking lots. From here, visitors proceed through side streets and town squares to the curved main streets and the open spaces.

The typologies of main street, side street, arcade, town square, ceremonial park, festival park, science park and nature park have been taken from the traditional city and transformed and abstracted. The resulting system provides a fixed circulation system with spaces and uses that are flexible and adaptable.

　　庆典中心是对这个地区性的商业步行街在功能、技术、美学和体验等方面进行重新思考的一次尝试，将都市购物的体验转移到城郊区域。地方购物中心通常体现出商品流通和空间管理的模式。与之相反，都市购物行为则是由商业、文化、娱乐和庆典等所组成的一个丰富而多样的混合物。

　　该中心的总体形象是一座由攀爬在一些构架上的葡萄树所形成的"绿色墙体"来围合而成的城市，通过巨大的"门"来赋予各种不同的购物区恰当的尺度感。"门"的后面就是停车库或地面停车场。从这里，游客穿过次要街道和市镇广场就可以到达弧形的主街和开放空间。

　　主街、次街、廊道、市镇广场、典礼园、节日园、科学园以及自然园在类型学上取自传统的城市，并将其进行了一些改变和抽象化。最后建成的体系形成了一种确定的流通系统，但空间和功能却是灵活多变，具有很好的适用性。

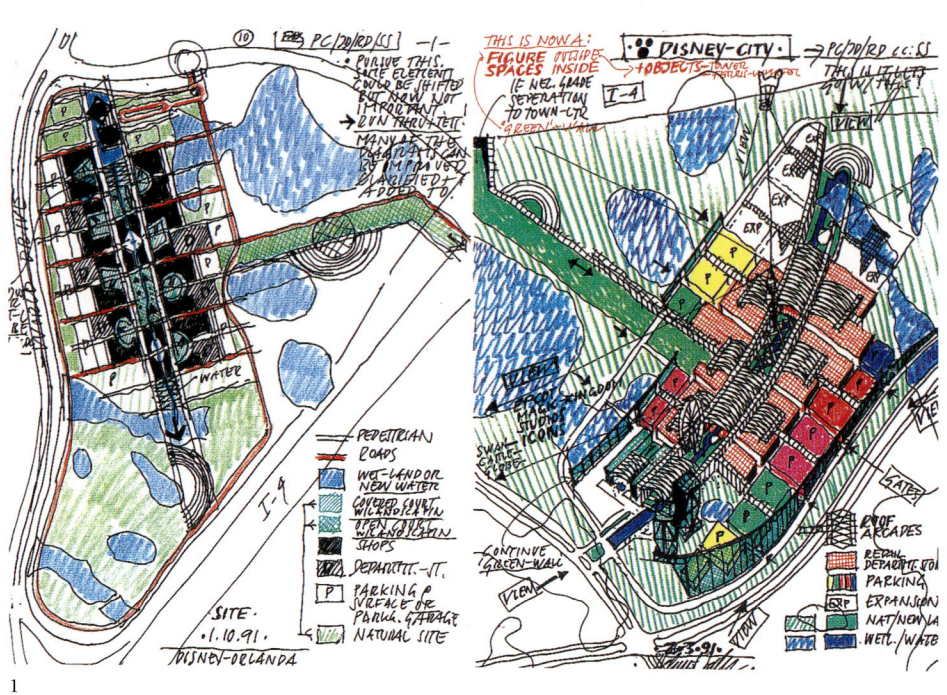

1

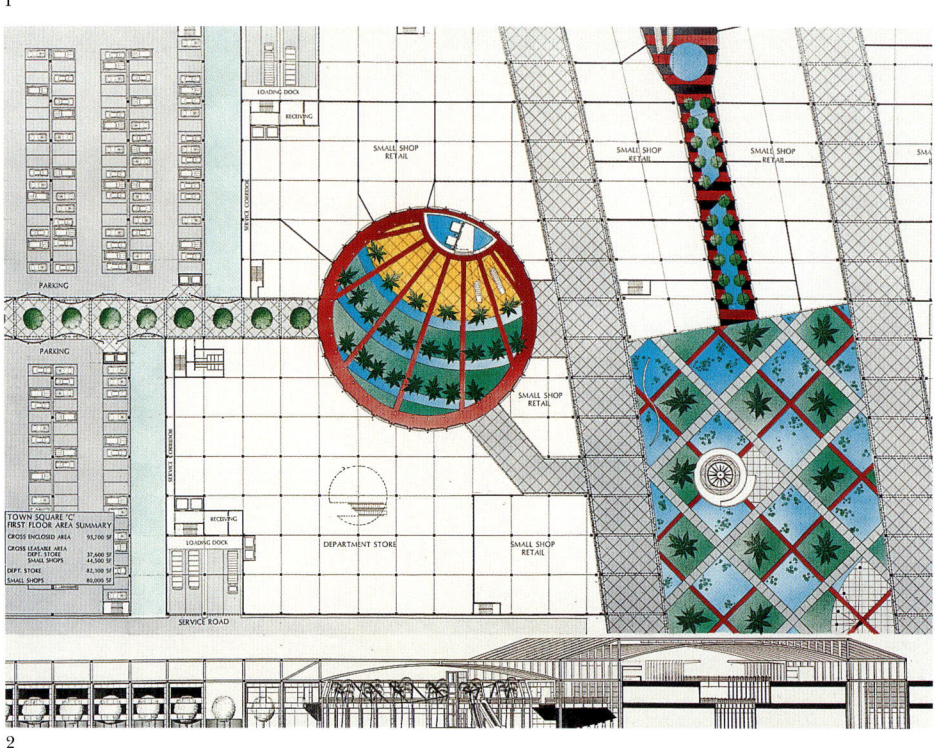

2

3

1 Concept sketch
2 Town squares
3 Enclosing "green wall" with frame and "gates"
4 Main street, arcade, town squares, retail and parking
5 Arcade
6 Nature park
7 "Green wall"
8 Festival park

1 构思草图
2 市镇广场
3 用框架和"门"来围合"绿色墙体"
4 主街、廊道、市镇广场、零售商店和停车场
5 廊道
6 自然园
7 "绿色墙体"
8 节日园

4

5

6

7

8

Navy Pier

Design 1991
Chicago, Illinois
Metropolitan Pier and Exhibition Authority
500,000 square feet
Steel frame
Glass infill, fabric and glass roof

海军码头

设计　　　1991
伊利诺伊州，芝加哥市
大都市码头和展览管理局
500,000平方英尺
钢框架结构
玻璃填充墙，织物和玻璃屋面

Navy Pier is not a single building, but a flexible arrangement of structures that allow a variety of different activities. Steel arcades are the connecting elements between the east and west wings. These contain the mechanical and electrical systems. Simple, arched steel girders can be spanned between the two buildings whenever space or a roofed area becomes necessary. The open spaces, bound by arcades, are left uncovered. This system allows for many closed, covered and open spaces, and guarantees the important continuity of the pier's profile.

From the outside, Navy Pier is a clear, linear object. Inside, it is a series of buildings, spaces and experiences: a linear city, which grows out into Lake Michigan. The pier landscape expands the public space of the lake-shore-park with a series of strong, visually exciting and useful public spaces.

　　海军码头不是一个单体建筑而是构造的灵活排列，在那能开展各种不同的活动。钢廊道是东西翼的连接要素。在这里容纳了机械和电力系统。无论何时需要空间或屋面遮盖的区域，则可在两栋建筑之间横跨起简单的拱形的钢桁支架。廊道限定的开放空间没有覆盖。这样的体系能允许许多围合的、覆盖的和开放的空间的共同存在，也确保了码头外形重要的连续性。

　　从外看，海军码头是一个清晰的线性的物体。在内部，它是一系列建筑、空间和经历的组合：一个线性的城市，向外扩展直至密歇根湖。码头景观通过一系列丰富的，视觉上令人激动的并且实用的公共空间来实现湖岸公园公共空间的延伸。

1

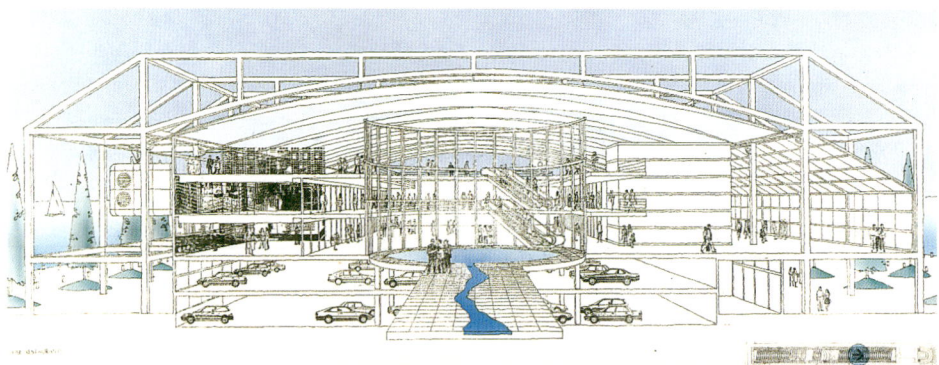

2

3

1　温室花园
2　餐厅
3　大观览车观看平台，在拱形钢桁支架区的开放的线性系列空间内

1　Winter garden
2　Restaurants
3　Ferris wheel and observation platform within open linear series of spaces with areas of arched steel girders

San Diego Convention Center

Design 1984
San Diego, California
San Diego Convention Bureau
Joint Venture Architects: Martinez/Wong Associates Inc.
690,000 square feet
Steel frame with long-span folded trusses
Glass curtain wall and metal panels

圣迭戈会议中心

设计　　　1984
加利福尼亚州，圣迭戈
圣迭戈会议局
合作建筑师事务所：马丁内斯/王联合有限公司
690,000 平方英尺
钢框架结构、大跨度折叠式桁架结构
玻璃幕墙和金属面板

The architect's foremost intention was to create a truly public building, devoid of the hermetically sealed character of most convention centers, which turn their back towards the city.

The entry functions and vertical movement corridors face the city and are revealed through glass walls. Public terraces face the water and are accessible by a ramp system from the boardwalk. Connecting the internal access with the public access is the Great Hall which serves as the entry to all meeting functions.

The architectural themes in this building are nautical images which can be identified with San Diego—lighthouses, masts, guy-wires, decks, bridges and cranes.

The footprint of the building is generated by the confines of the site. The forms of the building's distinct parts are enhanced by surface treatment and ornament, generated by structural and constructional necessities. Thus the building recalls much of the glass and steel architecture of the great exhibition buildings of the past.

建筑师的主要意图是创作一个真正具有公共性特征的建筑物，它没有大多数会议中心所具有的密封式的特征，那些建筑是背对着城市的。

入口功能区和垂直活动走廊面对着城市，并通过玻璃幕墙显现出来。公共活动平台面朝水面，可通过一个坡道系统从木板散步道到达这些平台。连接内部入口和公共入口的是一个大厅，它成为通向所有会议功能区的入口。

这栋大楼里的建筑主题是航海形象，如灯塔、桅杆、拉索、甲板、船桥和升降架等，从而人们可以从中看出圣迭戈的特征。

这栋建筑的平面形式是由建筑场地的红线所决定的。建筑的一些独特部分的形式则是通过表面的处理和装饰来予以强化，这些处理和装饰是由结构和构造的必要性所产生的。因而这栋建筑让人回想起过去那些杰出的展览建筑上体现出来的玻璃和钢结构建筑风格。

1

2

1　Lighthouses are a strong nautical image identified with San Diego
2　The Great Hall incorporates masts, bridges and cranes

1　灯塔是圣迭戈特有的强烈的航海形象
2　大厅里融入了桅杆、船桥和升降架的形象

Wisconsin Residence

Design/Completion 1980/1982
Eagle River, Wisconsin
2,500 square feet
Wood frame
Insulated glass and wood panels

威斯康星住宅

设计/竣工　1980/1982
威斯康星州，伊格尔河
2,500平方英尺
木框架结构
隔热玻璃和木墙板

The house, for a family of three, is planned as a vacation home, with the understanding that it may become a year-round residence at a later date. The building provides indoor living space and outdoor terraces on three levels. Also planned are an outdoor screened pavilion located near the water and a small marina. The site comprises 3.5 acres of lakefront property in northern Wisconsin. The house is located in the center of a wooded slope, affording views of the lake and opposite shore.

With its entry bridge, stair tower and lake pavilion, the house establishes a strong geometric presence on this lakeshore site. The structural and symbolic gridded shapes present themselves in sharp focus through the surrounding cover. Further reinforcing the (ideal object) quality of these forms is their sequential composition, designed to be a descending procession from hilltop to shore. The house itself is elevated in the center of this procession by a classically inspired cornice.

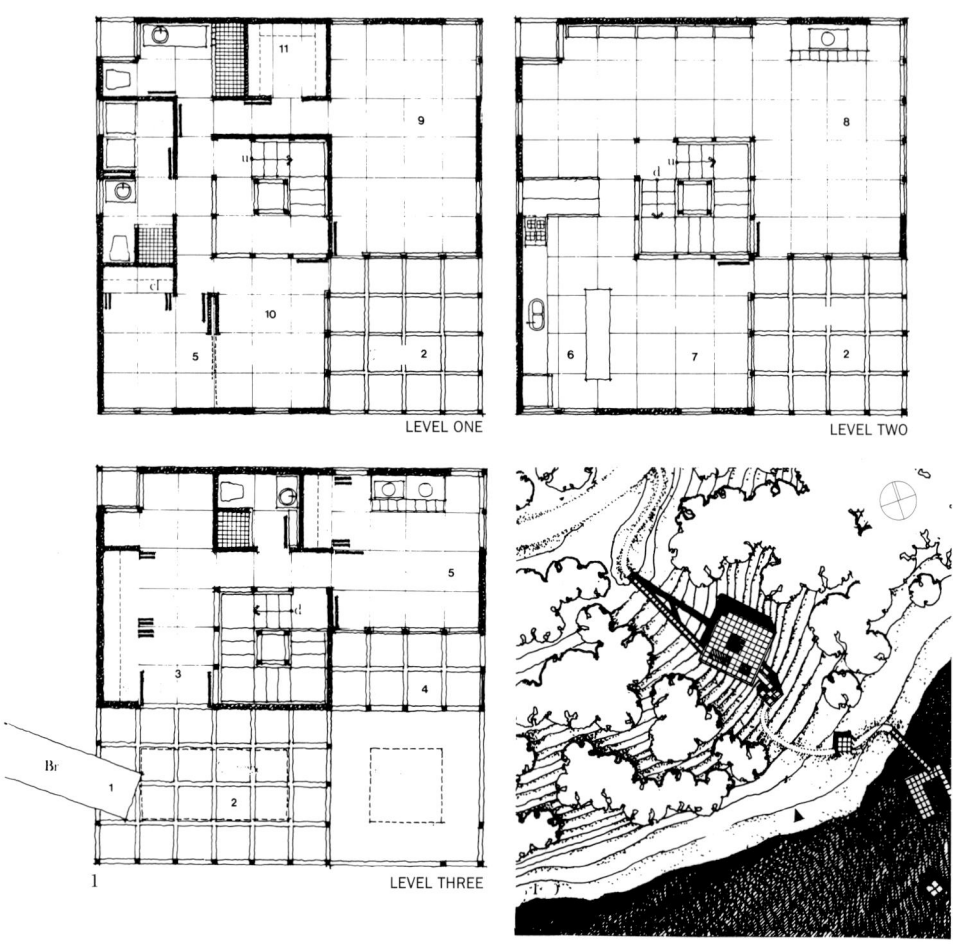

这栋住宅是为一个三口之家设计的度假别墅，但在设计的同时也考虑了它将来有可能成为主人全年居住的地方。这栋建筑形成了在三个不同标高上的室内居住空间和室外平台。该设计还包括了位于水边和小型码头的一个室外带屏风的亭子。建筑基地位于威斯康星州北部湖滨地带的一块3.5英亩大的土地上。这栋住宅坐落在一个树木繁盛的山坡的中心地段，从那可以看到湖面景色和对面的湖岸。

与入口天桥、楼梯塔和湖畔亭子一起，这栋住宅在这个湖滨基地上形成了一种强烈的几何表现。结构性和象征性的栅格形体让人能够透过周围茂密的树木而迅速地注意到它们。而其连续的构图则进一步强化了这些形式的理想品质，它被设计成从山顶到湖滨层层下降的一个序列。而住宅本身则通过一个具有古典风格的檐口架在这个序列的中央。

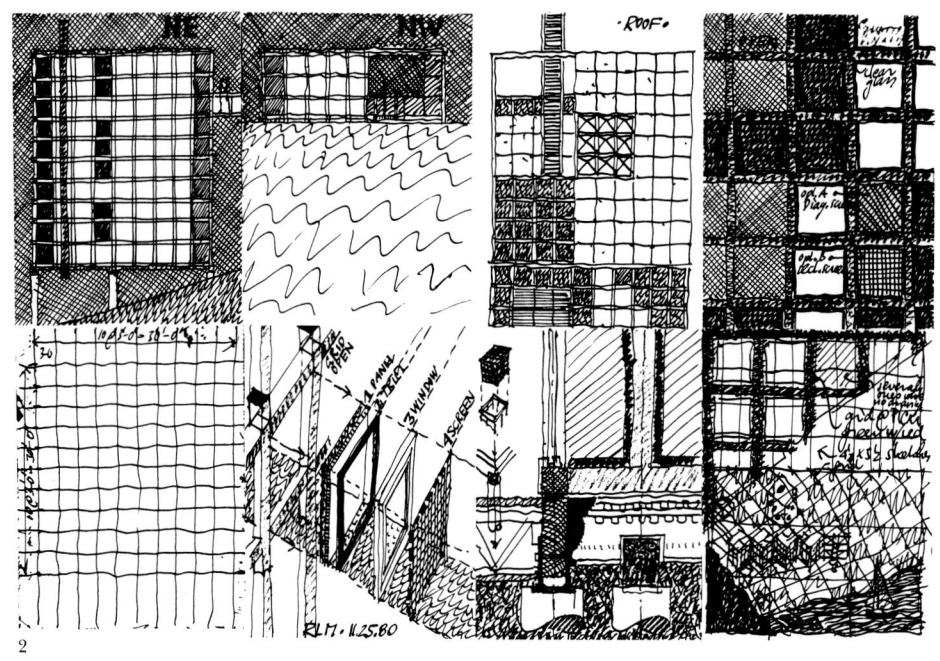

1–2 Helmut Jahn sketches
3 Night lighting of cube form
4 Lake elevation
5 Terrace entry upper level

1-2 赫尔穆特·扬的设计草图
3 立方体住宅的夜景
4 沿湖立面
5 平台入口最上层

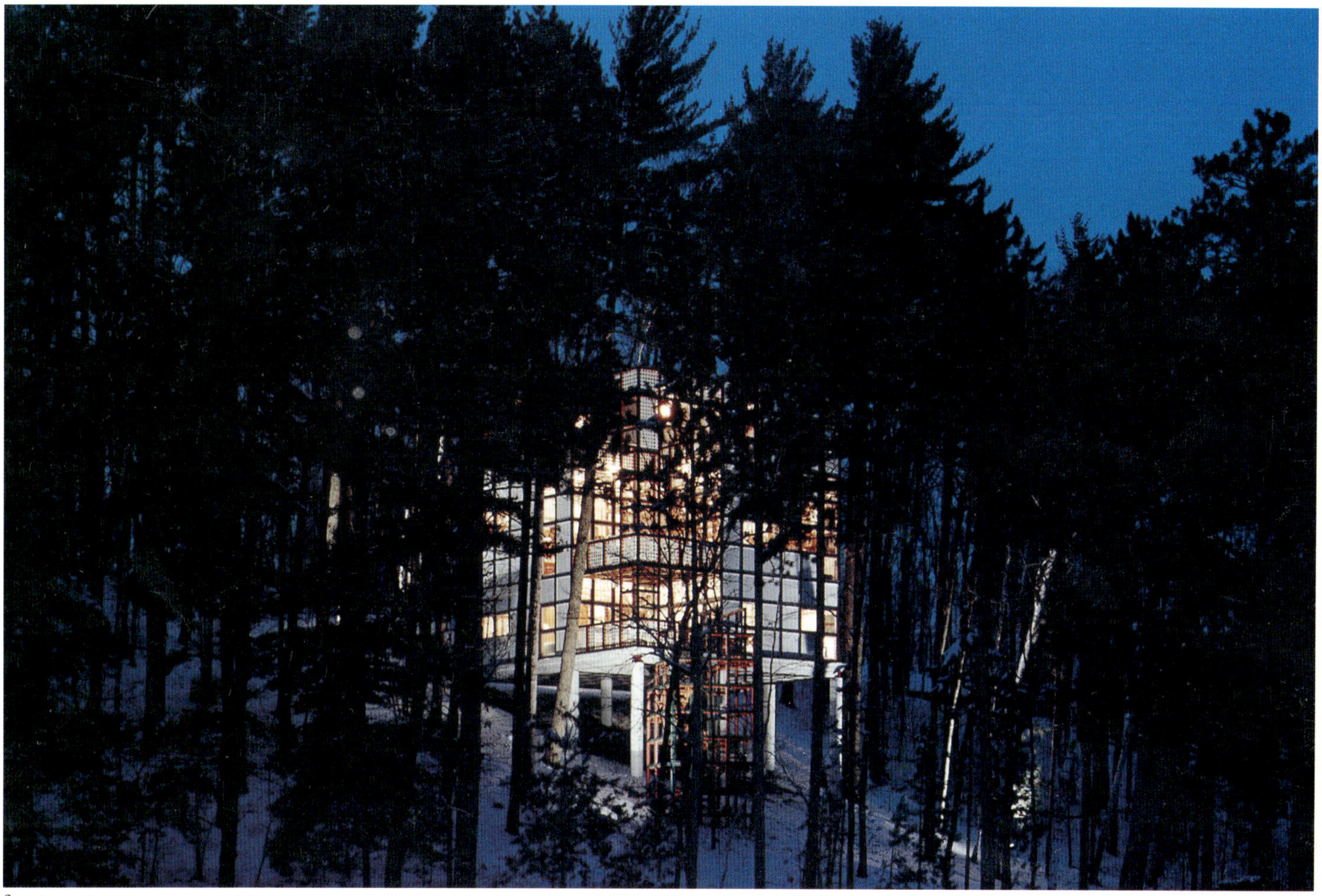

3

4

5

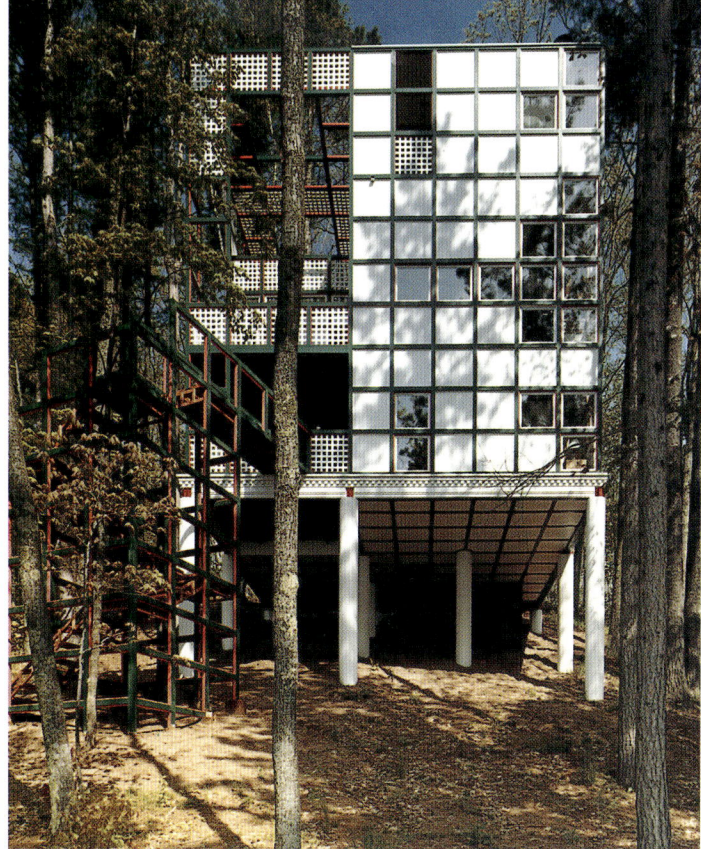

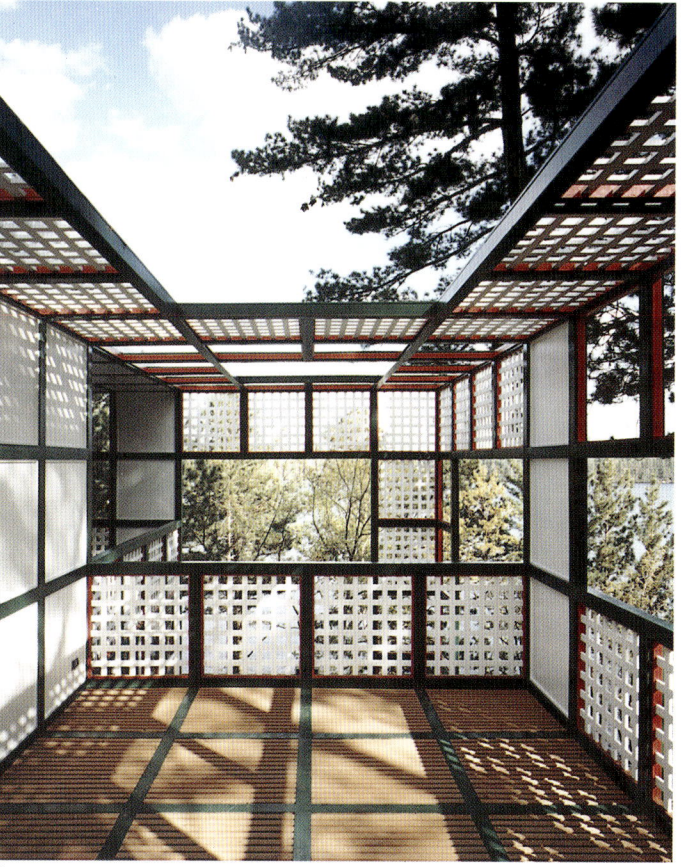

6 Side elevation with entrance bridge
7 Overall view with stair to lakefront
8 Terrace
9 Living area
10 Vertical stair core
11 Kitchen/eating areas

6 住宅及入口天桥侧立面
7 朝湖滨方向的住宅及楼梯全景
8 平台
9 起居空间
10 垂直的楼梯核心
11 厨房/就餐区

9

10

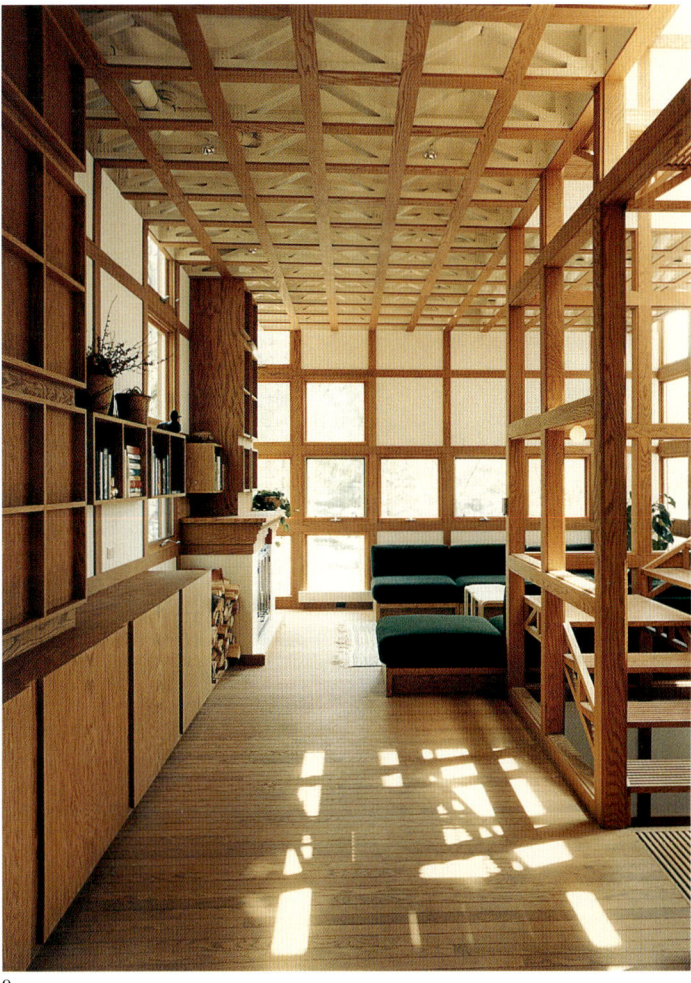

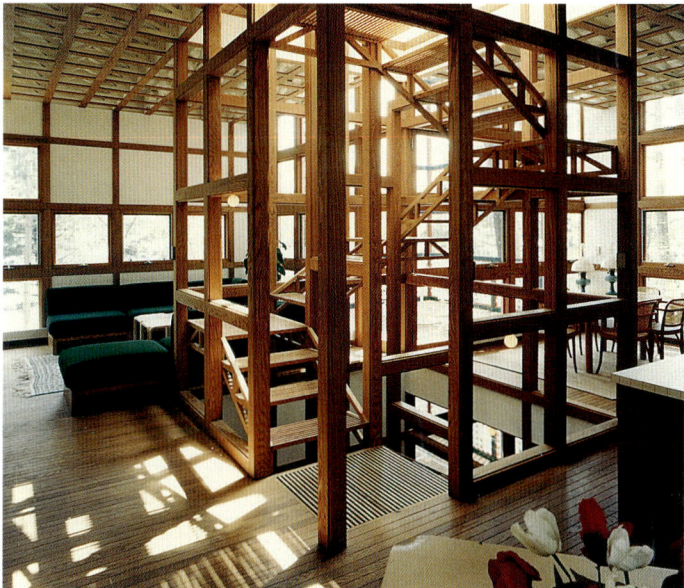

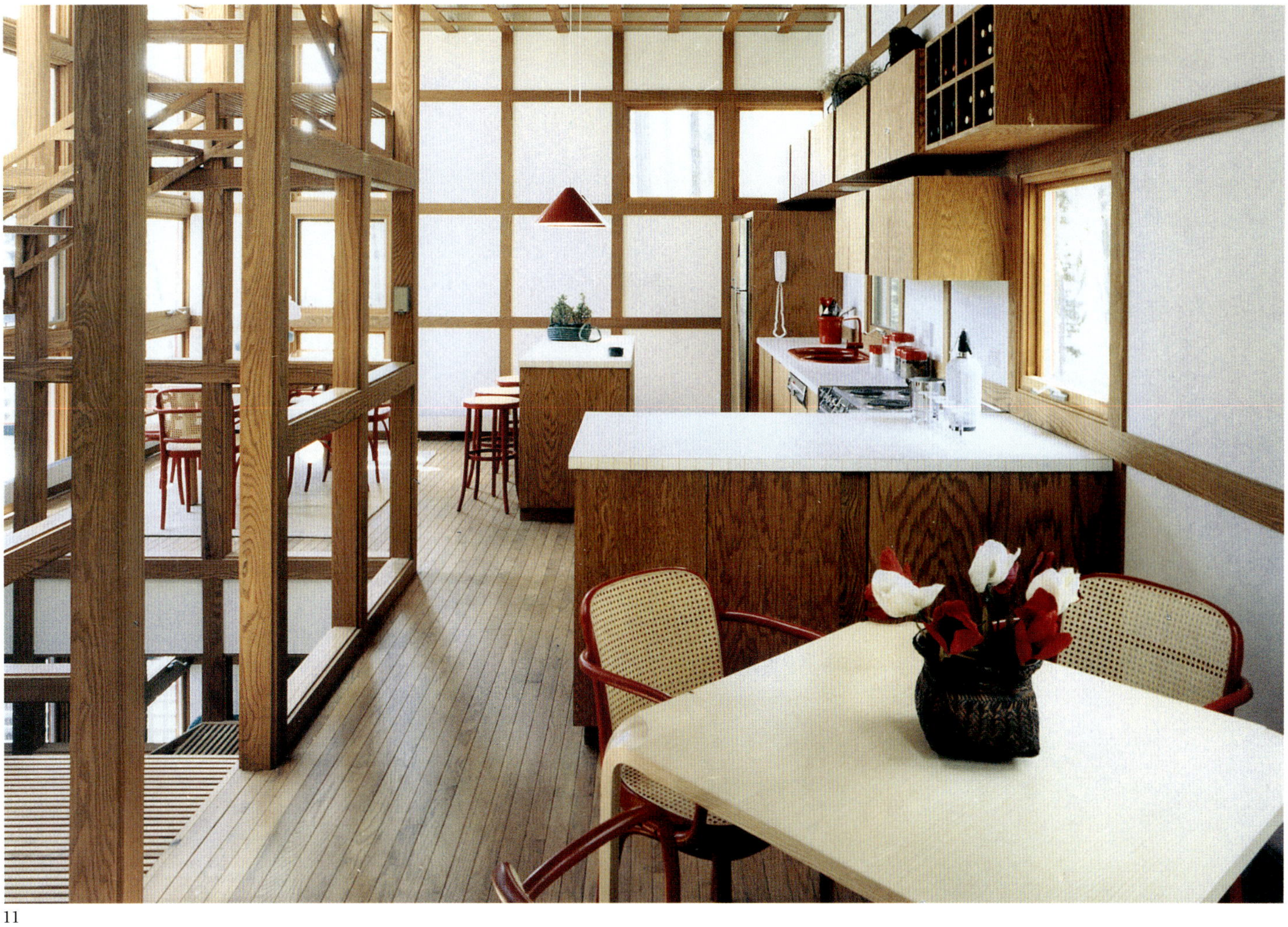

11

ANL/DOE Program Support Facility

Design/Completion 1978/1981
Argonne, Illinois
US Department of Energy
215,000 square feet
Steel frame
Clear anodized aluminum panels and sun shades, green painted mullions, silver painted structure and extensive skylighting

阿尔贡国立实验室/国家能源部项目辅助研究室

设计/竣工　1978/1981
伊利诺伊州，阿尔贡
美国能源部
215,000平方英尺
钢框架结构
透明阳极电镀铝板以及遮阳板，绿色油漆竖框
银色油漆结构及大型采光天窗

The program called for office and support space for the Chicago Branch of the Department of Energy and the Argonne National Laboratories. The brief envisages the building as a pilot demonstration project for passive solar energy use, the use of solar collectors, and energy-conscious design. The design disrupts the natural wooded setting of the site as little as possible. The round shape was generated in response to the road network, the non-directional character of the site, and (on a symbolic level) by the association it evokes with the sun as an energy source. A segment of the circle is removed to provide south orientation for the location of the future solar collectors. These collectors will step back to provide exterior shading for the glazed area. The amount of western exposure is minimized to reduce peak air-conditioning loads.

The round shape is completed by a retention pond which increases reflectivity for the solar collectors and daylight entering the building.

　　该项目为能源部芝加哥分部和阿尔贡国立实验室提供办公室和辅助空间。设计任务书将这栋建筑设想为一个在利用被动式太阳能、使用太阳能收集器和具有能源意识的试点示范项目。该设计尽可能地减少对建筑场地的自然森林环境的破坏。建筑的圆形外观与道路系统和建筑场地的无方向性特征相呼应，并且（在象征意义的层次上）通过联想，它使人想起太阳是一个能量的来源。圆形的一部分被去掉，使建筑留出朝南的面，以便为将来安装太阳能收集器提供位置。这些太阳能收集器将以退台的方式来安置，从而为安装玻璃的部分提供室外遮阳。该设计尽量减少建筑在西向的暴露面，以降低空调负荷的峰值。

　　圆形的外观通过一个保留的池塘来补充完整，而这个池塘不但为太阳能收集器增加了对阳光的反射，也让更多的光线进入到建筑之中。

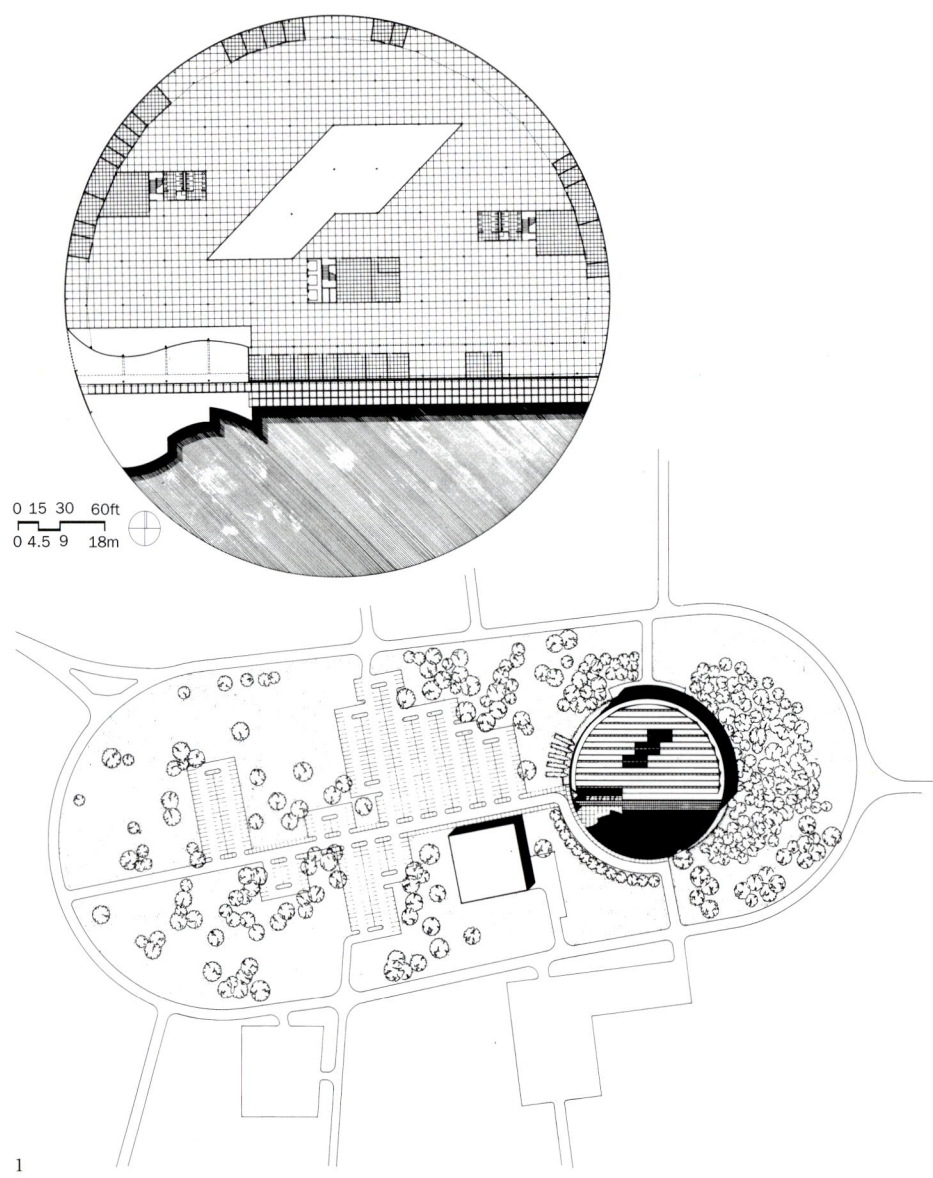

1

2

1 Third-floor plan and site plan
2 South-facing facade, sun shading and entrance
3 Curved side wall
4 Entrance
5 Vestibule
6 Light shafts maximizing natural daylight

1 第三层平面图和总平面图
2 南向立面，遮阳板及入口
3 曲线型的侧墙
4 入口
5 门廊
6 采光竖井使尽可能多的自然光进入建筑

3

4

5

6

ANL/DOE Program Support Facility 29

Area 2 Police Headquarters

Design/Completion 1977/1982
Chicago, Illinois
135,000 square feet
Steel frame
Blue insulated metal panels with panels of clear and frosted insulating glass block

二号区警察局总部

设计/竣工　　1977/1982
伊利诺伊州，芝加哥市
135,000 平方英尺
钢框架结构
蓝色隔热金属面板及透明和磨砂隔热玻璃砖

This government building serves a multitude of functions: area police center, district police center, police detention center, courts complex, sheriff's security complex, and vehicle maintenance facility.

The varied functions generate a range of configurations within a 660-foot long and 170-foot wide continuous building. The location of the building on the site creates zones for public and staff parking.

The public entry is through an open courtyard, which can also be used for community-related gatherings. This entry court ties into the circulation system which is clear and comprehensible, and provides the necessary separation of public and private activities.

Wall materials were chosen to provide as much natural light as possible, creating an openness not often found in this building type. All walls on the building's long sides are of glass block with bands of clear insulating glass where uses demand outside orientation. The court is completely glazed, and the end walls are of insulated metal panels.

1

2

3

这栋政府大楼有许多功能：地区警察中心、地方警察中心、警察拘留中心、法院综合楼、县治安综合大楼、以及车辆维护设施。

在660英尺长、170英尺宽的连续的建筑中，多样化的功能产生了一系列的构成关系。这栋建筑在场地上的位置形成了一些用于公共和员工停车的地带。

公共入口要通过一个开放式的庭院，它也可用作与社区相关的集会活动。这个入口庭院被结合在流线系统之中，显得既清晰又易于理解，并且形成了公共活动和私密活动之间所必需的隔离。

在墙体材料的选择上力争最大限度地为大楼提供自然采光，从而形成了一种在此类建筑中少有的开敞性特点。沿建筑长边的所有墙体都采用了玻璃砖和透明隔热玻璃带，因为这里的功能要求具有外向型的特征。庭院的四周都使用了玻璃，而山墙面则采用了经过隔热处理的金属面板。

1 Section with central court and flying structure	1 中央庭院和飞翼结构的剖面图
2 Street facade and public access	2 沿街立面与公共入口
3 Blue metal panel with infill of clear and frosted glass block	3 蓝色金属面板，填充墙为透明和磨砂玻璃砖
4 Internal stair	4 室内楼梯
5 Lobby	5 门厅

De La Garza Career Center

Design/Completion 1976/1980
East Chicago, Indiana
City of East Chicago
106,000 square feet
Steel frame
Press-formed steel sash, insulated panels and gasket glass

德拉加扎职业中心

设计/竣工　1976/1980
印第安纳州，东芝加哥
东芝加哥城
106,000平方英尺
钢框架结构
压型钢格窗，隔热面板和密封玻璃

A vocational high school building containing shops, classrooms and support space, the center is sited to separate public access from services and parking. Shops are located in an 80-foot-wide high-bay area; classrooms and support spaces are in an 80-foot-wide, two-story tract with 40-foot bays.

This arrangement allows for the clear linear organization of functions, the efficient use of space, and the consolidation of major program elements. The organization is expressed at the short ends where the structure is left exposed, revealing the building section and also providing for future expansion in that direction.

Standard systems are used for speed of construction, quality control and cost benefits. The structure is of truss, girders and joists on pipe columns; the exterior wall of press-formed steel sash and insulated panels and gasketed glass; interior walls are masonry.

这栋职业高中大楼里包括了车间、教室及辅助空间。职业中心的选址是为了将公共通道与设备房和停车场分开。车间位于80英尺宽的大开间区域，教室和辅助空间位于一个80英尺宽、两层高的大楼内，其开间为40英尺。

这种布局方式可以对功能进行明晰的线性组织，使空间得以有效地使用，并且能够强化主要的项目要素。建筑短边的结构被暴露出来，从这里可以看到对这种组织方式的表达。在这个地方揭示出建筑的剖面关系，也方便在这个方向上进行将来的扩展。

为了提高施工速度、进行质量控制和提高成本效益，该建筑使用了标准体系。结构采用在管状立柱上的桁架、大梁及托梁相组合的形式，外墙则采用了压型钢格窗、经过隔热处理的面板和密封玻璃，而内墙为砌块。

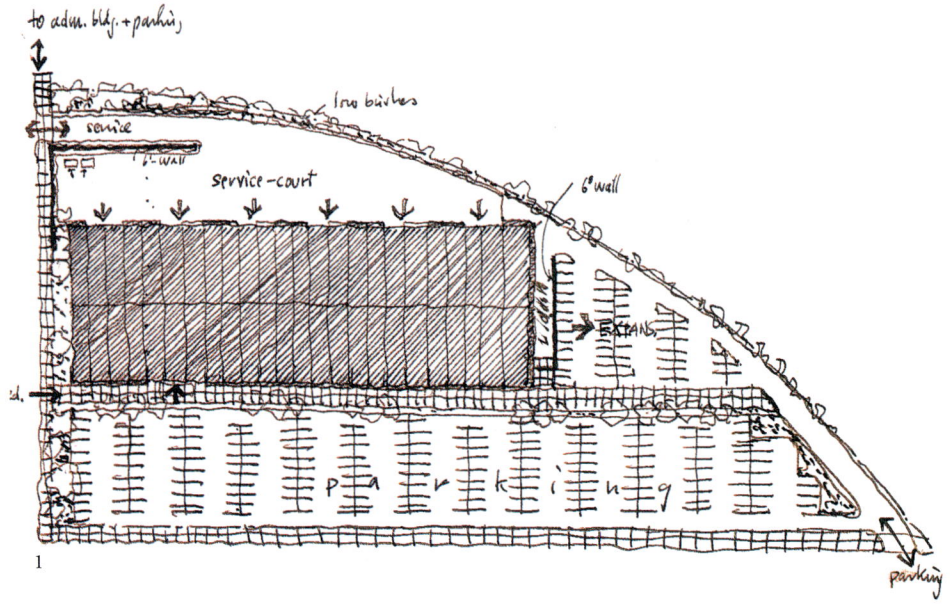

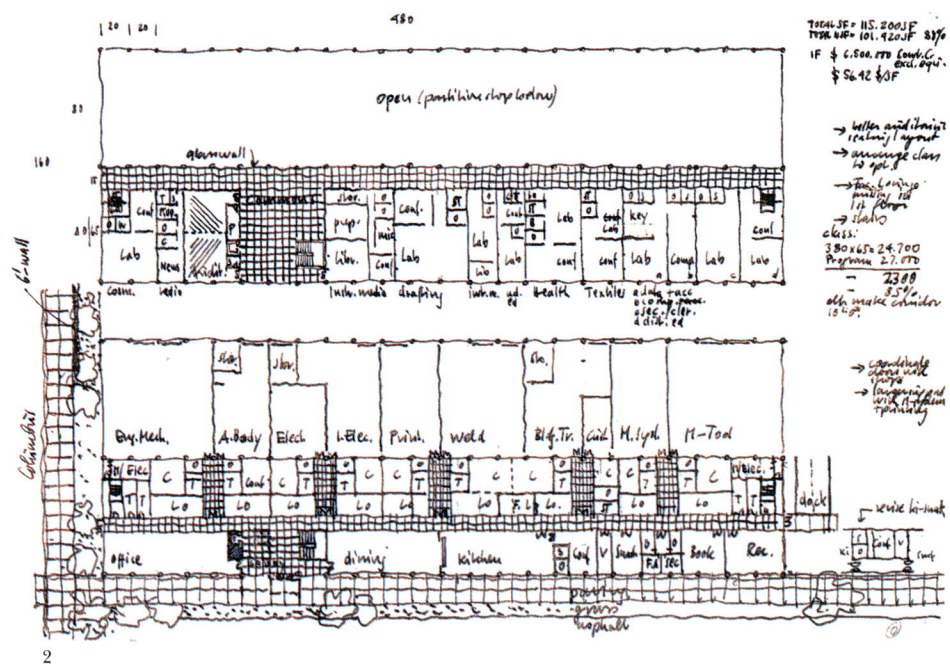

1　Sketch: site plan
2　Sketch: circulation, classroom and shop area
3　End wall elevation with mechanical unit
4　Green structure/yellow enclosure
5　Entrance
6　Shop entrance
7　Two-story shop area with expressed building systems

1　草图：总平面图
2　草图：流线关系、教室和车间区
3　带有机械设备的山墙立面
4　绿色结构/黄色围护体
5　入口
6　车间入口
7　两层高的车间区，表现出建筑体系

3

4

5

6

7

De La Garza Career Center 33

Rust-Oleum Corporation Headquarters

Design/Completion 1976/1978
112,000 square feet (includes half of parking area below building)
Vernon Hills, Illinois
Rust-Oleum Corporation
90,000 square feet
Exposed steel frame with metal deck
Alternate layers of glass and white enameled metal panels

拉斯特—欧勒姆公司总部

设计/竣工　　1976/1978
112,000 平方英尺（包括建筑下停车库的一半面积）
伊利诺伊州，弗农希尔斯
拉斯特—欧勒姆公司
90,000 平方英尺
暴露式钢框架结构及金属楼面
玻璃和白色瓷釉涂层金属面板复合材料

The building explores ideas of circulation and movement, the use of natural light, and the creation of a pleasant, humane working environment.

In order to preserve the available site area for landscaping, the structure is raised above the covered parking area that provides space for 112 cars. Employees enter from the parking level through secured lobbies, while visitors enter through level 1.

A central organizing spine is the three-dimensional circulation axis, and contains all building services. The skylight space creates openness and orientation and separates the office space into four flexible and modular areas which can be parceled according to changing requirements.

The exterior walls are completely glazed to strengthen the inside–outside relationship. Alternating horizontal bands of glass and aluminium reinforce the horizontal character of the building.

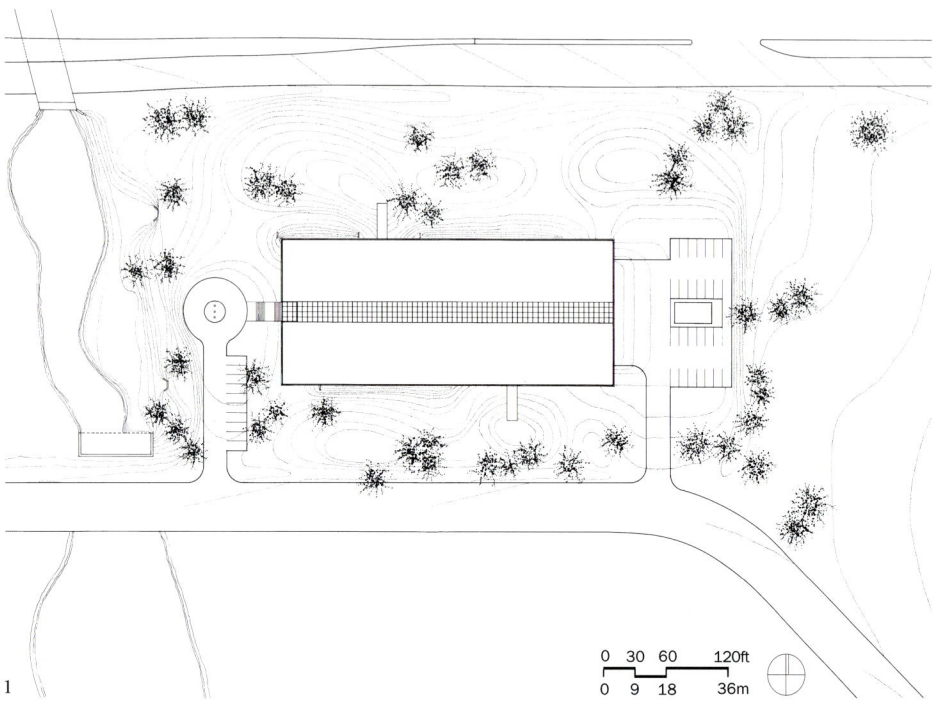

该建筑在流线和活动方式、对自然光线的利用以及创造一个愉快而充满人性的工作环境等方面的概念上进行了探索。

为了将该建筑场地留出来用于景观建设，建筑被架空起来，下面是能容纳112辆车的室内停车库。工作人员通过安全大厅从停车库层进入，而访客则从一层进入建筑之中。

一条集中布置的主干线成为其三维的流线交通轴，其中也包括了所有的建筑设备用房。通过天窗来采光的空间部分给人以开敞感和方向感，并且将办公空间划分为四个灵活的模数式区域，这些空间可以根据不断变化的需要而进行组合。

外墙全部采用玻璃幕墙，从而加强建筑内外的联系。交错出现的水平式的玻璃和铝材饰带强化了建筑水平的特征。

1　Site plan
2　Front elevation
3　Elevation sketch
4　Floor plan/circulation sketch
5　Circulation axis

1　总平面图
2　正立面
3　立面草图
4　楼层平面/流线设计草图
5　流线交通轴

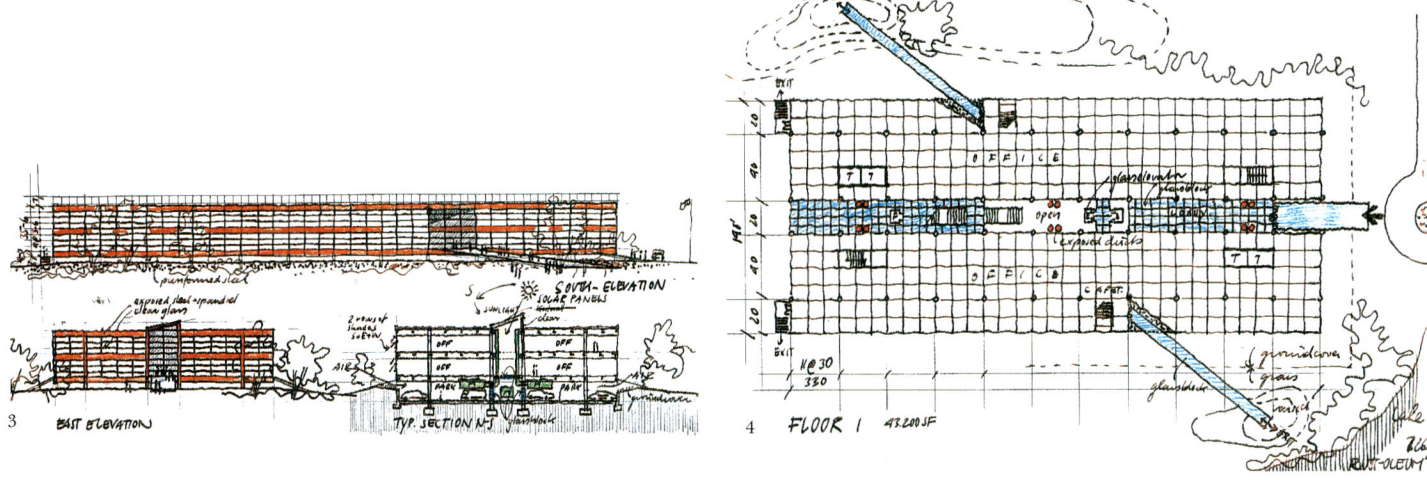

Rust-Oleum Corporation Headquarters 35

Saint Mary's Athletic Facility

Design/Completion 1976/1977
South Bend, Indiana
St Mary's College
42,000 square feet
Steel frame with long-span trusses
Translucent insulated fiberglass panels and a curved clerestory

圣玛丽体育运动馆

设计/竣工　1976/1977
印第安纳州，南本德
圣玛丽学院
42,000 平方英尺
钢框架及大跨度桁架结构
透明隔热玻璃纤维板和曲线式天窗

The primary use of the structure is as an athletics facility; however, it can also function as an exhibit or assembly facility for graduations, concerts, or social events. It contains two flexible multi-purpose areas, two racquetball courts, lockers, faculty offices, student lounge, lobby, and common support and utility space. Bleacher seating accommodates 1,800 people for spectator sports.

The racquetball courts and support spaces, which are fixed in size and use, are located on two floors at one end of the building. Adjacent to this is the high-bay multi-purpose area for tennis, volleyball and basketball. Gymnastics, dance and fencing are accommodated in a low-bay area at the other end of the structure.

Two circulation trunks along the perimeter of the upper level connect all functional elements, and establish the order and organization for expansion.

该建筑的最初用途是作为一个体育运动设施，然而，它也可作为一个用于有关毕业典礼、音乐会或社会活动的展览或集会建筑。该建筑内有两个可以灵活变化的多功能区、两个壁球场、更衣室、教师办公室、学生休息室、大厅和公共辅助和设备空间。露天看台座位能容纳1,800名观众来观看体育比赛。

壁球运动所需要的场地和辅助空间在尺寸和功能方面都是固定的，这两个球场位于建筑一个端头上的两个楼层中。靠近它们的是大开间的多功能区，用于网球、排球和篮球运动。体操、舞蹈和击剑场所位于建筑另一端的一个小开间区域内。

环绕着建筑上层周边展开的两条交通主线将所有功能元素连接起来，也为将来的扩展建立了秩序和组织方式。

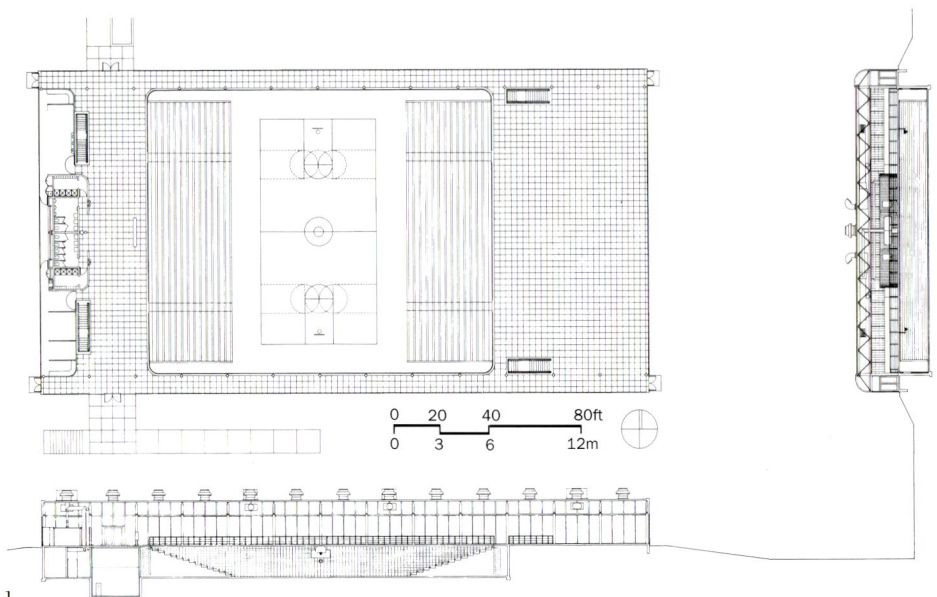

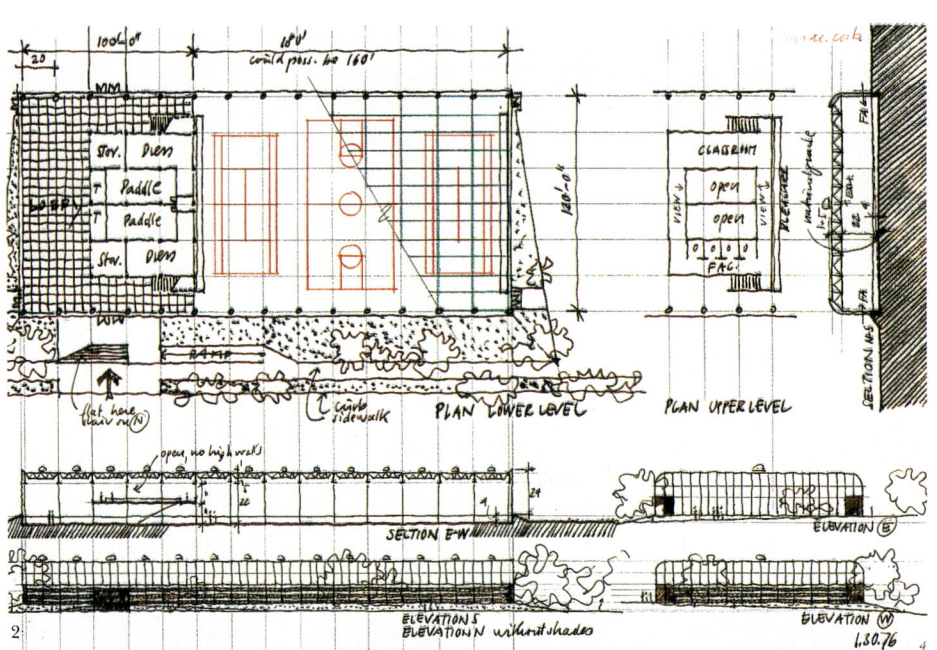

1 Floor plan/sections
2 Concept sketches
3 Front elevation
4 Playing field with spectators area
5 Playing field with clerestory
6 Integrated mechanical system
7 Multi-purpose area/playing field
8 Entrance/circulation trunk/clerestory

1 楼层平面图/剖面图
2 构思草图
3 正立面
4 带观众区的运动场地
5 带天窗的运动场地
6 综合机械系统
7 多功能区/运动场地
8 入口/交通主线/天窗

4

5

6

7

8

Saint Mary's Athletic Facility 37

Michigan City Public Library

Design/Completion 1974/1977
Michigan City, Indiana
Public Library System
35,000 square feet
Exposed steel frame with steel joists
Translucent insulated fiberglass panels with north-oriented clerestory

密歇根城市公共图书馆

设计/竣工　1974/1977
印第安纳州，密歇根市
公共图书馆体系
35,000平方英尺
暴露式钢框架及钢托梁结构
透明隔热纤维玻璃板与北向天窗

The building serves as the main branch facility for the Michigan City Public Library system. It occupies a downtown site, creating a spatial closure to Franklin Mall, the main commercial spine.

The building contains areas for adult services, reference and information, children's services, audiovisual resources, technical services, administration, and a business office. Common areas include the lobby, circulation desk and a flexible meeting room complex.

The program is organized within a one-story "loft space". An interior court is located asymmetrically along the diagonal axis. All office and work areas are located along the perimeter enclosed by free-standing partitions. Only the permanent and movable walls of the meeting room complex extend to the structure and roof to provide acoustical privacy.

　　该建筑是密歇根市公共图书馆系统中的主要机构。它位于城中心地段，对富兰克林购物街这个主要的商业中心形成一种在空间上的围合。
　　建筑内包括成人服务区、会议和信息区、儿童服务区、视听材料区、技术设备区、行政区以及一个业务办公室。公共区域包括大厅、流通柜台区和一个多功能综合会议室。
　　所有的活动都被组织在一个单层的"阁楼式空间"（有回廊或夹层的大空间——译者注）之中。一个室内庭院不对称地位于对角轴线上。所有的办公和工作区都布置在建筑的周边，外墙为独立式隔墙。只有综合会议室的固定和活动墙体一直延伸到结构和屋面，用以实现良好的声学效果。

1　平面图
2　带天窗的入口
3　周边的交通走廊
4　透明的外墙
5　室内景观庭院

1　Plan
2　Entrance with clerestory
3　Perimeter circulation corridor
4　Translucent exterior wall
5　The interior landscaped courtyard

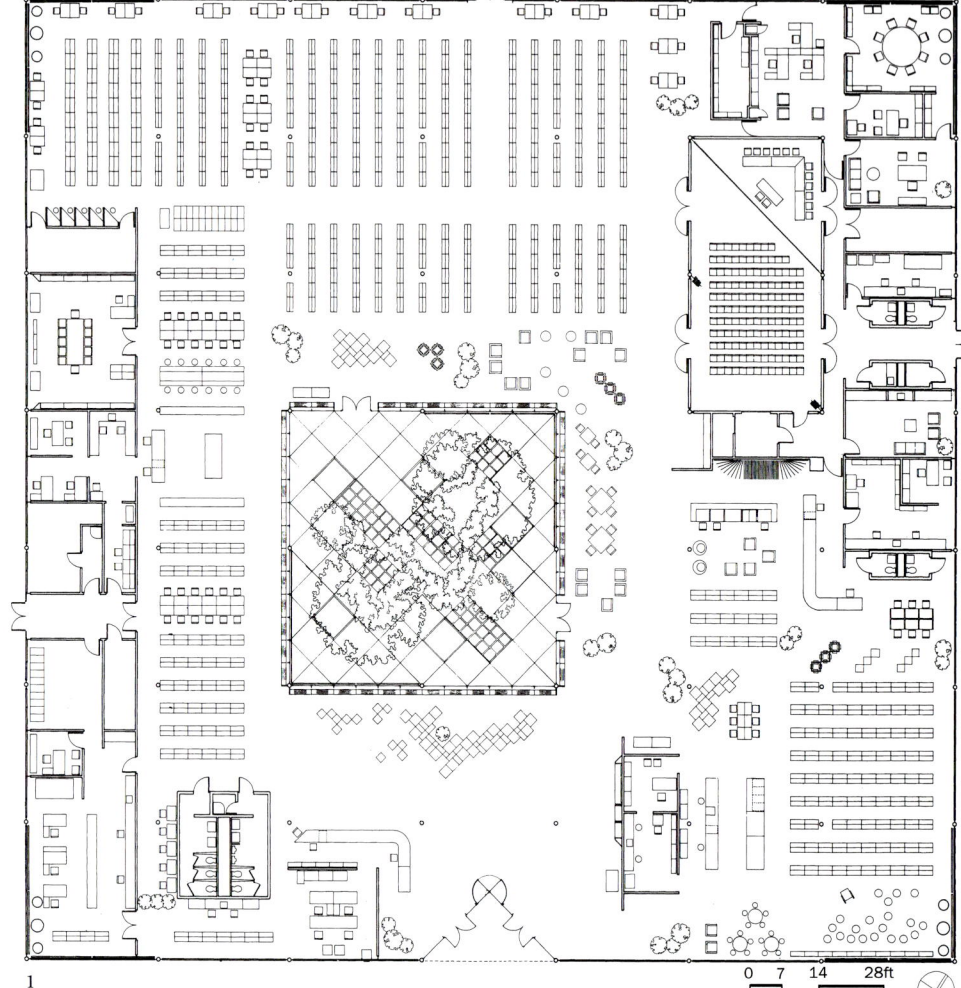

3

4

5

Michigan City Public Library 39

Auraria Library

Design/Completion 1974/1976
Denver, Colorado
Auraria Higher Education District
184,000 square feet
One-way, post-tensioned cast-in-place concrete; reinforced concrete
Aluminum skin of fixed glass, operating sash and insulated metal panels

奥拉利亚图书馆

设计/竣工　1974/1976
科罗拉多州，丹佛市
奥拉利亚高等教育区
184,000平方英尺
单向后张式现浇混凝土，钢筋混凝土结构
固定玻璃铝包皮、可开启窗扇及隔热金属板

The building serves as the central Learning Resources Center for a new college in Denver, and contains library, media production and media education facilities. The organization of the building into two levels above grade and a partial basement reflects its different functional elements: the media production center requires no natural light, while the reading and support space of the library require maximum exposure.

The two floors are developed as flexible "loft space". Two open courts placed asymmetrically within the plan subdivide each floor into various "use" areas, accommodate outdoor reading, and provide natural light and outside air. Three open stairways connect the two levels. Except for a set core on the two levels, all individual rooms are placed along the perimeter.

该建筑是在丹佛的一个新学院的主要学习资源中心，其中包括图书馆、媒体制作及媒体教育设施。设计时将该建筑布置在地面上的两个标高上并有部分地下室，以此来反映出建筑中不同的功能性元素：媒体制作中心不需要自然采光，而图书馆的阅读和辅助空间则要求获得尽可能多的自然采光。

建筑的两层楼都被设计成具有灵活性的"阁楼式空间"。两个开敞的庭院不对称地布置在平面内，它们再将每一层楼划分为不同的"功能"区。在庭院中设置了室外阅览场所，可以让人们享受自然光线和户外空气。三个开敞式的楼梯将两个楼层联系起来。除了在两个楼层上的固定核心区外，所有的房间都沿外墙来布置。

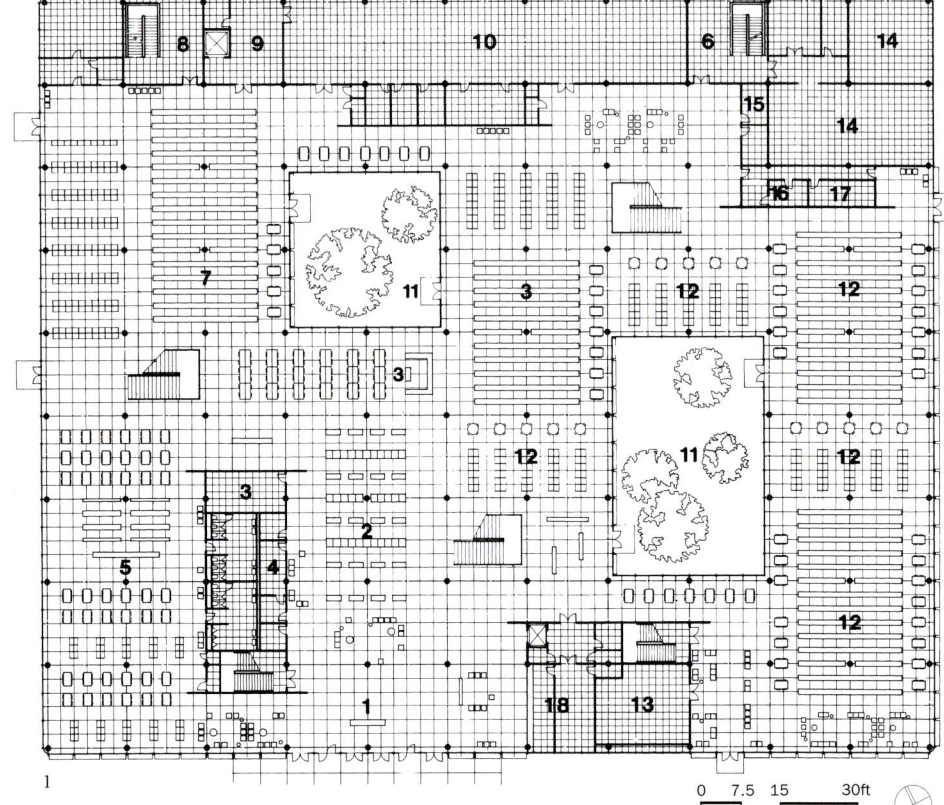

1. Entry/exit lobby
2. Public catalog
3. Reference
4. Inter-library loan
5. Reserve
6. Media education entrance
7. Periodicals
8. Vestibule
9. Shipping
10. Technical services
11. Outdoor reading
12. General
13. Mechanical
14. TV studio
15. Conference
16. Duplication
17. Typing
18. Circulation

1. 入口
2. 公共目录
3. 参考书目
4. 图书馆内部借书处
5. 储藏室
6. 媒体教育区入口
7. 期刊杂志处
8. 门廊
9. 运送处
10. 技术服务区
11. 室外阅览区
12. 普通阅览处
13. 机械设备处
14. 电视工作室
15. 会议室
16. 复印室
17. 打印室
18. 流通处

1　平面图
2　带遮阳板的正立面
3　带遮阳板的玻璃幕墙

1　Plan
2　Front elevation with sun shades
3　Glass wall with sun shades

Abu Dhabi Conference Center City

Design 1976
United Arab Emirates
Steel truss operable room with concrete substructure
Punched metal panels, sun shades and laminated glass

阿布扎比会议中心城

设计　1976
阿拉伯联合酋长国
钢桁架结构活动式房间，混凝土基础
打孔金属板、阳光板及夹层玻璃

This prize-winning concept developed by C.F. Murphy Associates in an international competition for a conference city in Abu Dhabi, United Arab Emirates, represents a unique approach to the problems of transportation, urban design, environmental impact, and energy conservation.

This city is to be located on a narrow peninsula in the Persian Gulf where the midday temperatures reach 120°F. The major facility on the site, the conference center, includes an assembly hall for delegation meetings, large exhibition spaces, restaurants, meeting rooms, and facilities for the world press. The plan also includes housing for delegates, visiting heads of State and other dignitaries and related support and service functions.

The main element in the organization of the city is the automated personal transit system which connects the major elements of the conference center, residences, service areas, and mode transfer points. The spatial concept is one of a complete climate-controlled environment between the circulation trunks.

这是 C·F·墨菲联合事务所在阿拉伯联合酋长国的阿布扎比的会议中心国际设计竞赛中的获奖方案，它提出了一种解决交通、城市设计、环境影响和节能等问题的独特方法。

这个会议中心城位于波斯湾中的一个狭窄的半岛上，那里中午的温度高达120华氏度。建筑场地中的主要设施是会议中心，它包括了一个用于代表团会议的大会堂、大型展览空间、餐厅、会议室和各国媒体所需要的设施。该方案也包括了供代表、来访国家首脑和其他重要人物下榻的地方以及相关的辅助和服务功能设施。

对这个城市进行组织的主要元素是自动化的人员运输系统，它将会议中心、住宅区、服务区和转车点等关键要素联系起来。在空间构思上将它设计成一个在流线主通道之间的全温控下的环境。

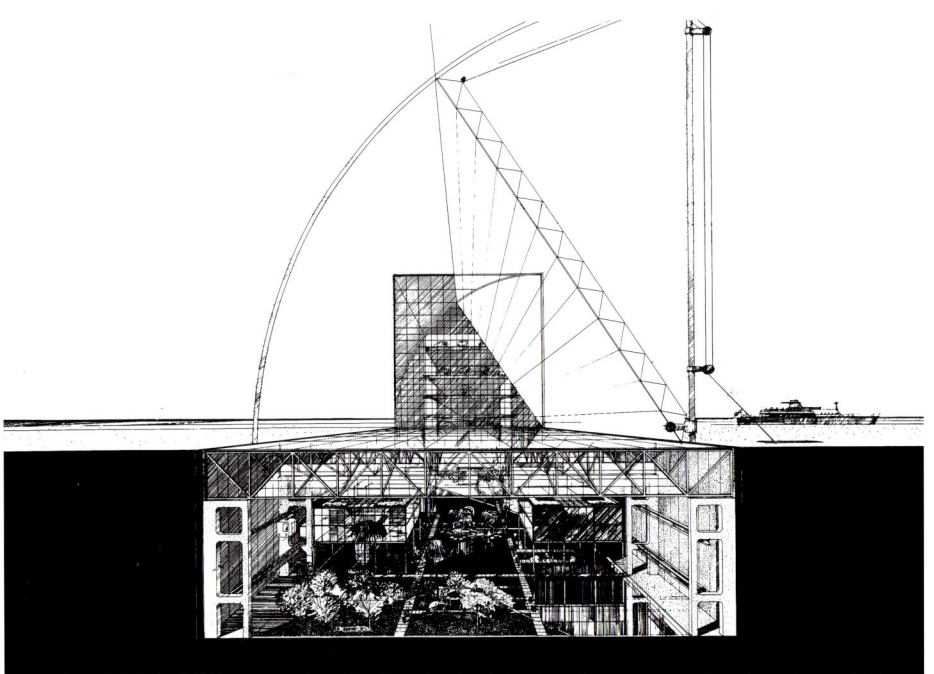

1

2

1　Section/perspective
2　Delegate housing with conference court in background and movable sun-shading roof

1　剖面图/透视图
2　代表们的住处，顶部为活动式遮阳屋面，背景上是会议中心庭院

Minnesota Government & History Center

Design 1976
St Paul, Minnesota
State of Minnesota
Cast-in-place reinforced concrete
Laminated glass-covered winter garden

明尼苏达州政府和历史中心

设计　　1976
明尼苏达州，圣保罗
明尼苏达州政府
现浇钢筋混凝土结构
夹层玻璃覆盖的温室花园

The building forms part of the existing Capitol complex, and houses the Minnesota Historical Society Museum as well as classroom space, public meeting rooms for legislative and state agency hearings, an auditorium, a public cafeteria and parking for 450 cars.

The building is completely below grade, allowing the landscaping to be the form-giving and directional element. This enhances and reinforces existing vistas and creates a more pedestrian-oriented precinct. A succession of spatial experiences leads to a large rectangular open space appropriate to the classical dignity and formality of the site.

The circulation system relates the public spaces in the building to a grand, glass-covered garden that extends most of the way between the State Office Building and the Historical Society Building. This space contains a garden restaurant and is open to all levels of the new building.

　　该建筑是现有州议会大厦综合楼的一部分，其中包括了明尼苏达历史社会博物馆、教室空间、供立法机构和州政府机构听证活动使用的公共会议室、一个礼堂、一个公共自助餐厅以及一个有450个车位的停车场。
　　该建筑完全处在地面标高以下，从而让景观成为赋予形态和方向性的要素。这种做法强化了现有的视景线，并形成了一个更加步行化的区域。一连串的空间体验将人引向一个巨大的矩形开放空间，该空间与建筑场地的庄重性和正式性很合适。
　　流线体系将建筑内的公共空间与一个巨大的玻璃花园联系起来，该花园向外延伸，占了从州政府办公楼到历史社会博物馆之间的道路长度的大部分。该空间包括了一个花园饭店，并且朝新建筑的各个楼层开放。

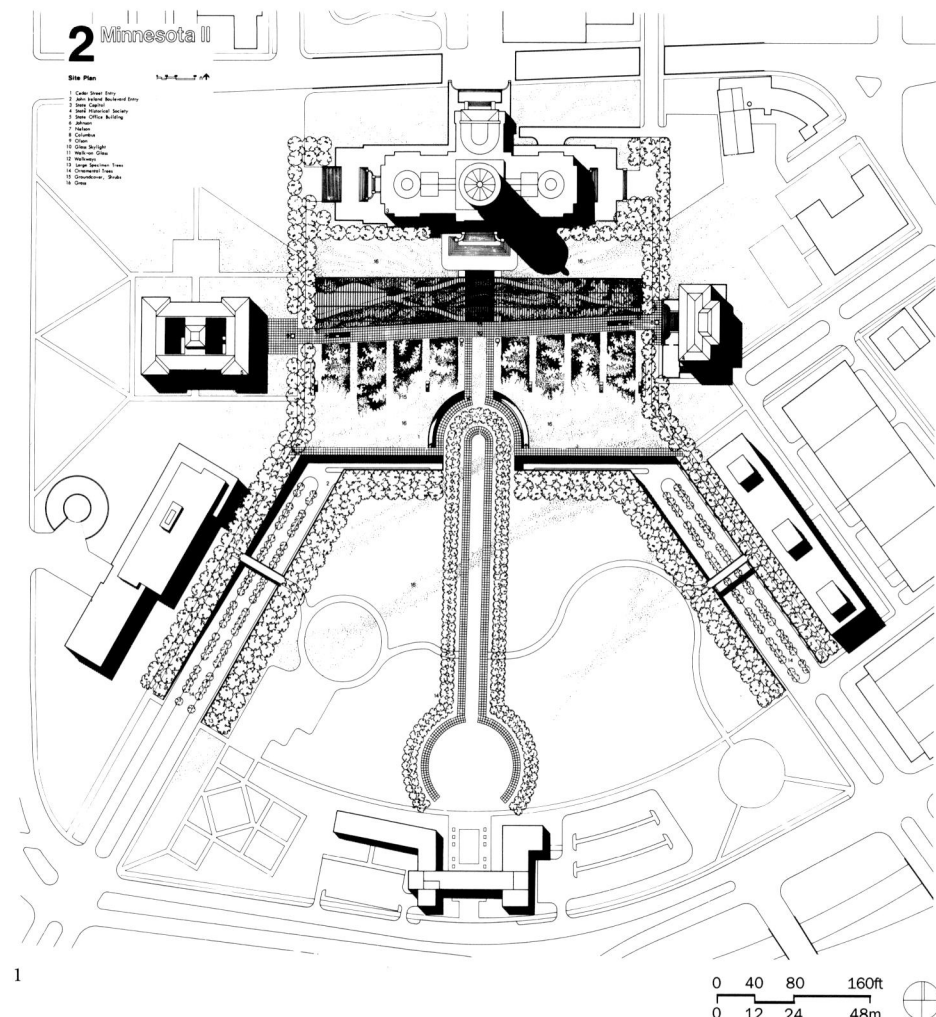

1　建筑总平面图
2　地下综合建筑物和玻璃花园，背景为州议会大厦

1　Site plan
2　Underground complex and glass-covered garden with Capitol building in the background

Kemper Arena

Design/Completion 1973/1974
Kansas City, Missouri
American Royal Arena Corporation
17,000 seats
Steel pipe trusses with secondary steel joists
Insulated metal panels

肯珀竞技场

设计/竣工　1973/1974
密苏里州，堪萨斯市
美国皇家竞技场公司
17,000 座
钢管桁架及次级钢托梁结构
隔热金属板

The brief called for a multi-purpose indoor arena seating 1,000 to 18,000 people to be used for hockey, basketball, and boxing; track events; musical and variety programs; horse and livestock shows; and conventions. The development occupies a former stockyards site in an industrial district undergoing redevelopment.

The building is a single-story structure with two major elements: the substructure (including the seating), and the superstructure and building enclosure. The substructure is of reinforced concrete. The seating in the upper tier is of L-shaped precast, prestressed concrete elements on cast-in-place bents. For the superstructure, the major subsystems of steel structure, steel decking and steel cladding panel system enabled the basic envelope to be completed in a short time.

The ground level of the structure is buried in an earth berm, which provides a walkway around the building at concourse level. The oval form of the concourse level transforms into a rectangle at the upper level through the addition of four mechanical rooms.

项目任务书要求设计一个有 1,000 到 18,000 个座位的多功能室内竞技场，可以用于开展曲棍球、篮球和拳击、径赛项目、音乐与各种活动，以及马匹和牲畜展览会，还能举办会议。该开发项目位于一个正在进行再开发建设的工业区内，该基地以前是一个牲畜围场。

该建筑为单层结构，有两个主要元素：下部结构（包括看台）、上部结构和建筑围合物。下部结构是由钢筋混凝土建成。在其上层部分的看台是在现浇排架上的 L 型预制预应力混凝土构件组成。上部结构分为钢结构、钢制支撑平台等主要子系统。钢面板系统使建筑的基本围护外墙结构可以在短时间内得以完成。

该建筑的地面标高被埋在一个土堤里，从而在中央大厅的标高上形成了一个环绕着该建筑的人行道。建筑在中央大厅标高上的椭圆形体处，通过增加四个设备用房，使其在上面一层变形为矩形。

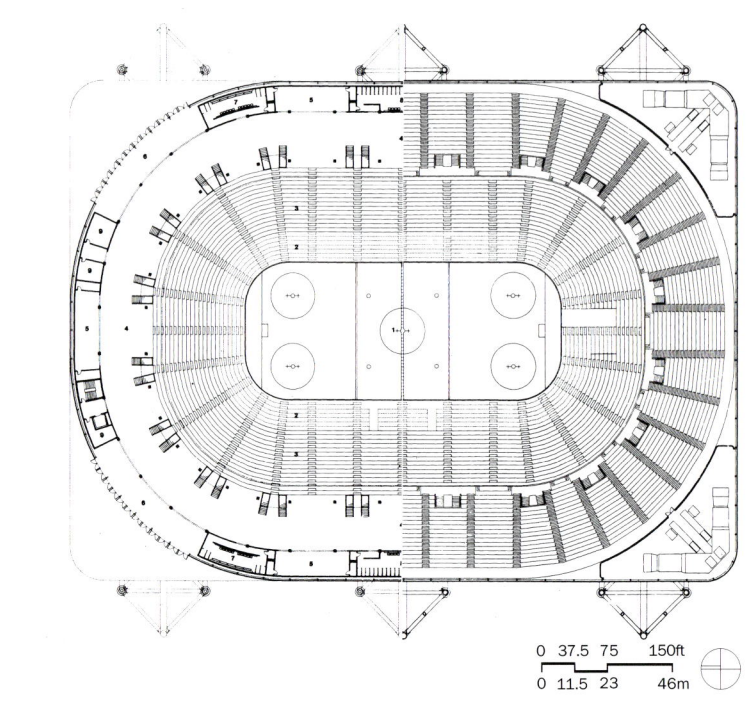

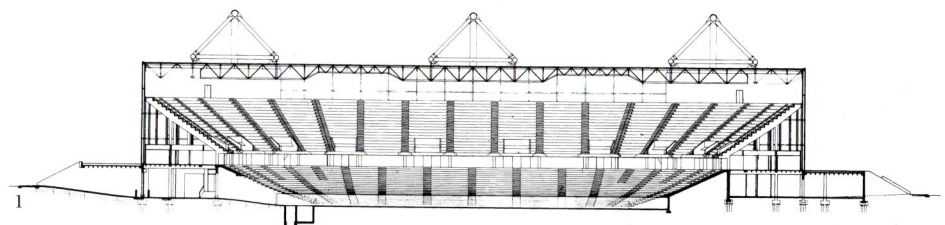

1

2

1 Lower and upper level seating plan/section
2 Interior arena seating
3 West elevation/section
4 West elevation
5 Rigid bent main exterior structure
6 Detail of tube elements

1 下层和上层看台平面图/剖面图
2 竞技场室内看台
3 西立面图/剖面图
4 西立面图
5 刚性排架形成的主要外部结构
6 管状要素的细部

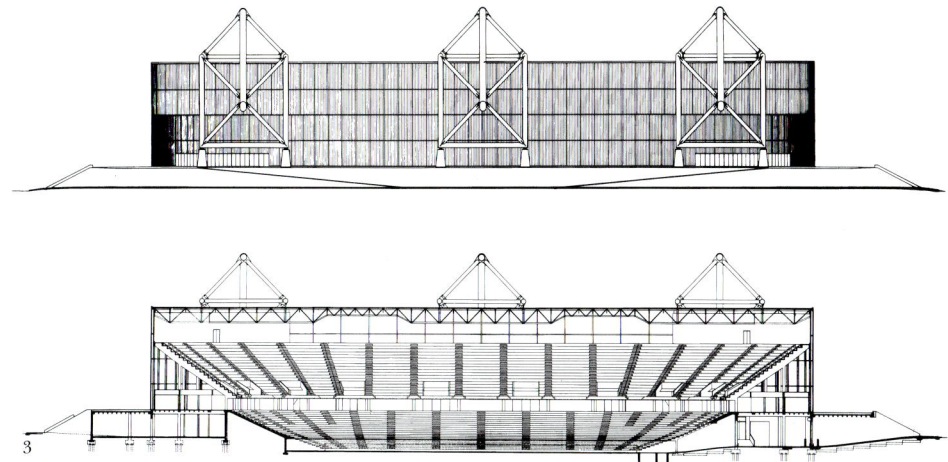

Kemper Arena 45

作品精选

Selected and Current Works

塔式建筑

- 48　雅加达通信塔楼
- 49　雅加达塔楼和昆宁干中心
- 50　21世纪塔楼
- 54　无穷尽发展的塔楼
- 55　东丘－涩谷塔楼
- 56　日立塔楼／卡尔泰克斯大楼
- 66　堪宁堡塔楼
- 67　IHZ塔楼
- 68　120北拉萨尔大楼
- 72　博览会塔楼／博览会大厅
- 80　喷泉广场西侧大楼
- 82　联邦航空局奥黑尔控制塔楼
- 83　吉隆坡一号城市中心
- 84　巴尔内特中心
- 86　横滨滨水地带MM21工程
- 88　东京特勒波特中心
- 90　莱克星顿大街425号
- 92　莱克星顿大街750号
- 94　城市之巅
- 95　J项目（1，2，3号）
- 96　威尔希尔／威斯特伍德
- 100　布里克尔大街1111号
- 101　北环城路第37号街区
- 102　奥克布鲁克台地的高层塔楼
- 104　自由城一号
- 108　西北中庭中心
- 112　西大街362号
- 114　公园大街塔楼
- 116　西地铁站
- 118　南渡口广场
- 119　哥伦布环道
- 120　西南银行塔楼
- 122　斜向大街第11号
- 124　芝加哥同业公会扩建工程
- 126　南瓦克一号
- 128　施乐中心

Tower Buildings

- 48　Jakarta Communications Tower
- 49　Jakarta Tower Kuningan Centre
- 50　21 Century Tower
- 54　Endless Towers
- 55　Tokyu-Shibuya Tower
- 56　Hitachi Tower/Caltex House
- 66　Fort Canning Tower
- 67　IHZ Tower
- 68　120 North LaSalle
- 72　Messe Tower/Messe Hall
- 80　Fountain Square West
- 82　FAA O'Hare Control Tower
- 83　Kuala Lumpur City Center 1
- 84　Barnett Center
- 86　Yokohama Waterfront MM21
- 88　Tokyo Teleport Town
- 90　425 Lexington Avenue
- 92　750 Lexington Avenue
- 94　Cityspire
- 95　Project J (1, 2, 3)
- 96　Wilshire/Westwood
- 100　1111 Brickell Avenue
- 101　The North Loop Block 37
- 102　Oakbrook Terrace Tower
- 104　One Liberty Place
- 108　Northwestern Atrium Center
- 112　362 West Street
- 114　Park Avenue Tower
- 116　Metro West
- 118　South Ferry Plaza
- 119　Columbus Circle
- 120　Bank of Southwest Tower
- 122　11 Diagonal Street
- 124　Chicago Board of Trade Addition
- 126　One South Wacker
- 128　Xerox Centre

Jakarta Communications Tower

Design 1995
Jakarta, Indonesia
PT Kuningan Persada
1,500,000 square feet
Braced steel frame
Aluminum frame with infill glass and aluminum louvers

雅加达通信塔楼

设计　　1995
印度尼西亚，雅加达市
昆宁干·佩萨达房地产集团
1,500,000 平方英尺
斜撑钢框架结构
铝材框架内填玻璃，铝百叶窗

This telecommunications tower for television, radio, satellite, and microwave transmissions will become a major national symbol of Indonesia.

Two alternatives were studied: a tower only, and a tower with an office building at its base. An amenity building with retail, convention, entertainment, and parking facilities surrounds the tower and links it to its urban setting.

The plan configuration is a tripod. The structure is a triangulated perimeter tube. The low aspect ratio of the structure and the braced perimeter system minimize steel weight and deflections. A glass skin covers the main faces of the triangle. The observation and restaurant levels project out from the tower in pinwheel fashion, permitting views of Jakarta, the sea, and the tower itself. A holographic artwork symbolizing Indonesia's five principles of philosophy crowns the top of the tower.

电视通信塔楼用于电视、电台、卫星和微波传输，它将成为印度尼西亚的一个重要的标志。

在设计过程中研究了两种可能性方案：一个方案是单体塔楼，而另一个方案由一座塔楼及其基部的办公楼构成。一座容纳了零售、会议、娱乐和停车设施的舒适的建筑围绕着这个塔楼，将它与城市环境联系起来。

该方案在构成上是一个三角架形状。其结构采用一个成三角形的边筒。该建筑的高宽比很小，周边体系采用斜撑，使建筑的重量和挠度都达到最小。玻璃面层覆盖了三角形结构的大部分表面。观景区和餐厅区的楼层以风车状方式从塔楼上伸出去，使人们可以观看雅加达、大海和塔楼本身。一件代表着马来西亚哲学中的五个原则的全息式艺术作品被安放在塔楼的顶上。

1　塔楼方案
2　塔楼/办公楼方案

1　Tower option
2　Tower/office option

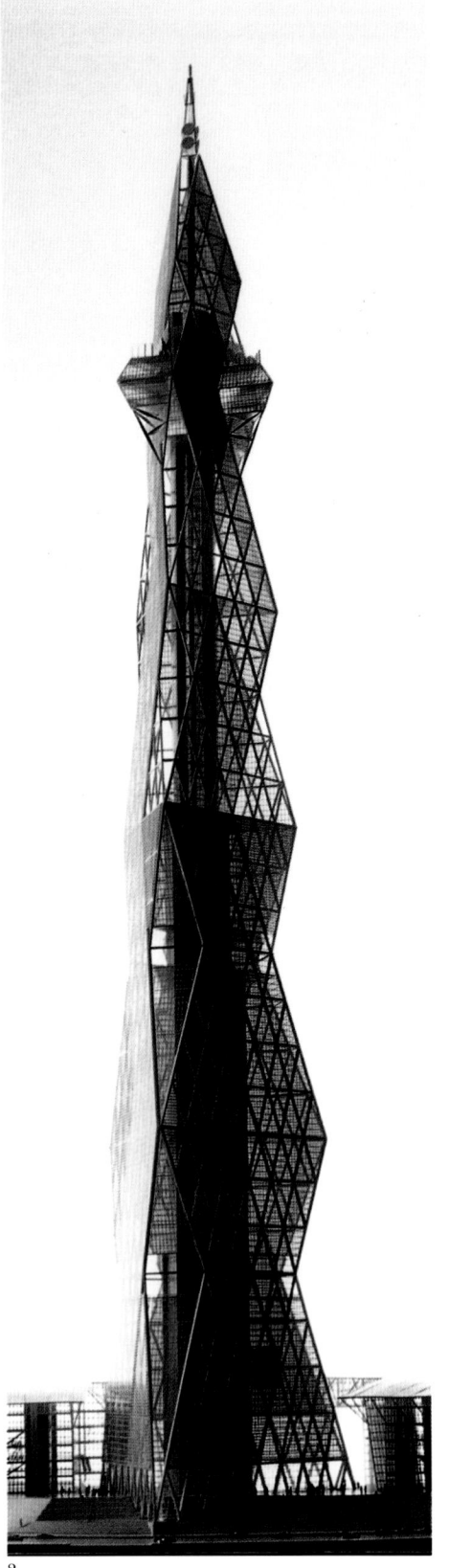

1
2

Jakarta Tower Parcel 8 Kuningan Center

Design 1994
Jakarta, Indonesia
Pacific Metrorealty
1,448,000 square feet
Steel frame with seismic superframe and modular gravity frame
Glass in aluminum mullion system, steel trellis, precast panels, stone faced panels and stone grid systems

雅加达塔楼和昆宁干中心

设计　　1994
印度尼西亚，雅加达市
太平洋米特罗蕾亚蒂公司
1,448,000 平方英尺
钢框架结构，防震超强框架和模数重力框架
铝材竖框体系中镶玻璃，钢格架，预制面板，石材饰面面板和石格栅系统

The design for Jakarta Tower combines the geometric simplicity of the square and circle with the beauty and dynamics of the spiral, representing progress and an ascendent future. It also represents a synergy of tropical garden and high-rise building.

The subtle spiral expressed in the curtain wall is articulated by set-back terraces with vine-covered trellis rising the full height of the building. A tinted soffit at the set-back areas insures a clear reading of the building motif from all angles.

The tower is located on a retail base which provides its urban context. A network of gallerias and arcades integrates the office, banking, and retail functions. The central retail atrium is defined by walkways of clear geometries which overlay a common structural grid. Canopies and sunscreens foster connections and linkages as well as insuring user comfort and flexibility of functions.

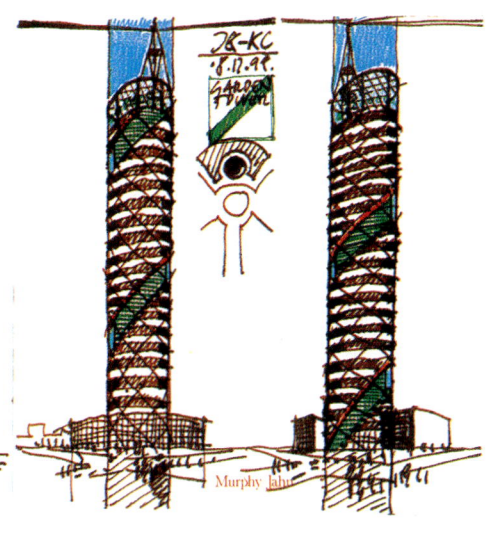

1

雅加达塔楼的这个设计结合了方形和圆形的几何简洁性，并有螺旋的美感和动感，象征着进步和一个上升的未来。它也代表了热带花园和高层建筑的有机结合。

这个在幕墙上体现的微妙的螺旋形式是通过后退的平台来构成的，而蔓生植物覆盖的格架一直延伸到建筑的顶部。一个在后退区的彩色的拱腹确保人们从各个角度都能清楚地读懂建筑的主题。

塔楼位于用于零售活动的基础之上，这个零售活动部分赋予了建筑都市的背景。一个由走廊和廊道的形成的网络将办公、银行和零售的功能统一起来。中央的零售中庭则被覆盖在一个普通结构栅格之上的有着清晰几何形体的人行走道衬托得格外分明。天篷和遮阳板促进了各个功能的联系和联结，也确保了使用者的舒适度和功能的灵活性。

2　　　　　　　　　　　　　　　　　3

1　Concept sketches
2　Spiral curtain wall with set-back vine-covered terraces
3　Entrance axis

1　构思草图
2　螺旋式幕墙以及后退的蔓生植物覆盖的平台
3　入口轴线

Jakarta Tower Parcel 8 Kuningan Center　49

21 Century Tower

Design/Completion 1993/1997
Shanghai, China
Everbright International Trust and Investment Corp.
Associate Architect: East China Architectural Design Institute
862,400 square feet
Steel frame
Painted red panels, blue and green glass with black frit pattern and gray frames

21世纪塔楼

设计/竣工　　1993/1997
中国，上海市
光大国际信托投资公司
合作建筑设计单位：华东建筑设计研究院
862,400 平方英尺
钢框架结构
红色面板，灰色框架蓝绿玻璃幕墙

21 Century Tower will occupy a unique site in the heart of the Lujiazui Finance and Trade Zone of the Pudong New Area, facing the Shanghai Bund.

The project consists of two components: the 49-story tower and the three-story podium building. Five nine-story skygardens form a third major element. These landscaped gardens are spaces to be used for informal meetings, meals or relaxation. Their structure is independent of the tower, and supports a glass facade which provides protection from the environment.

The structural principle employed for the tower is that of a braced tube. The structural system eliminates the column at the north-east corner which fronts the east–west axis, marking the ceremonial entrance and identifying the skygarden locations. The podium building, at the south-west corner of the site, is treated as a pavilion that is separated programmatically from the tower, but is tied to it by a system of canopies that provide covered passage.

21世纪塔楼位于浦东新区陆家嘴金融和贸易区的一个独特的地段上，其对岸就是上海外滩。

该工程由两部分组成：一座49层高的塔楼和一个3层高的裙房。5个9层高的空中花园构成了第三种主要的元素。这些景观花园中是用于非正式的会议、就餐或休闲的空间。空中花园的结构是独立于塔楼的，它支撑着玻璃幕墙，将建筑空间与外部环境分隔开来。

该塔楼采用了斜撑筒体的结构原理。该结构体系取消了在东西轴线端头上的东北角的柱子，从而突出了建筑的正式入口并标识出空中花园的位置。在基地的西南角的裙房被处理成一个轻质的附属建筑，在设计上将它与塔楼分隔开来，但又通过一系列的天篷将其与塔楼联系在一起，提供了一个有遮蔽的通道。

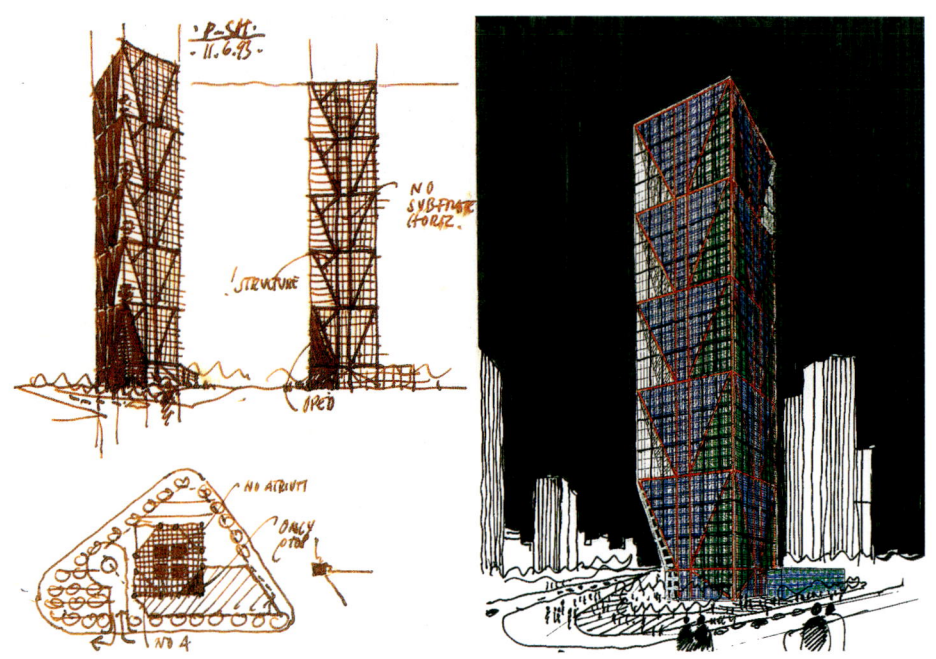

1

2

1 Concept sketches
2 Skygarden
3 Five nine-story skygardens
4 The axis reinforces the master plan's ceremonial axis
5 A singular object expressing transparency, color and structure

1 构思草图
2 空中花园
3 5个9层通高的空中花园
4 该轴线强化了总平面图上的形式轴线
5 一个单一性的物体表达了透明性、色彩和结构

3

4

5

21 Century Tower 51

6	Blue and green glass reinforces the structural pattern	6	蓝色和绿色强化了结构性图案
7	Skygardens are glazed in clear glass with stucture in red	7	空中花园由透明玻璃来围合，结构部分涂成红色
8	Ceremonial entrance and three-story podium detail	8	形式上的入口及3层高的裙房部分的细部
9	The structure is that of a braced tube	9	结构采用斜撑筒体体系

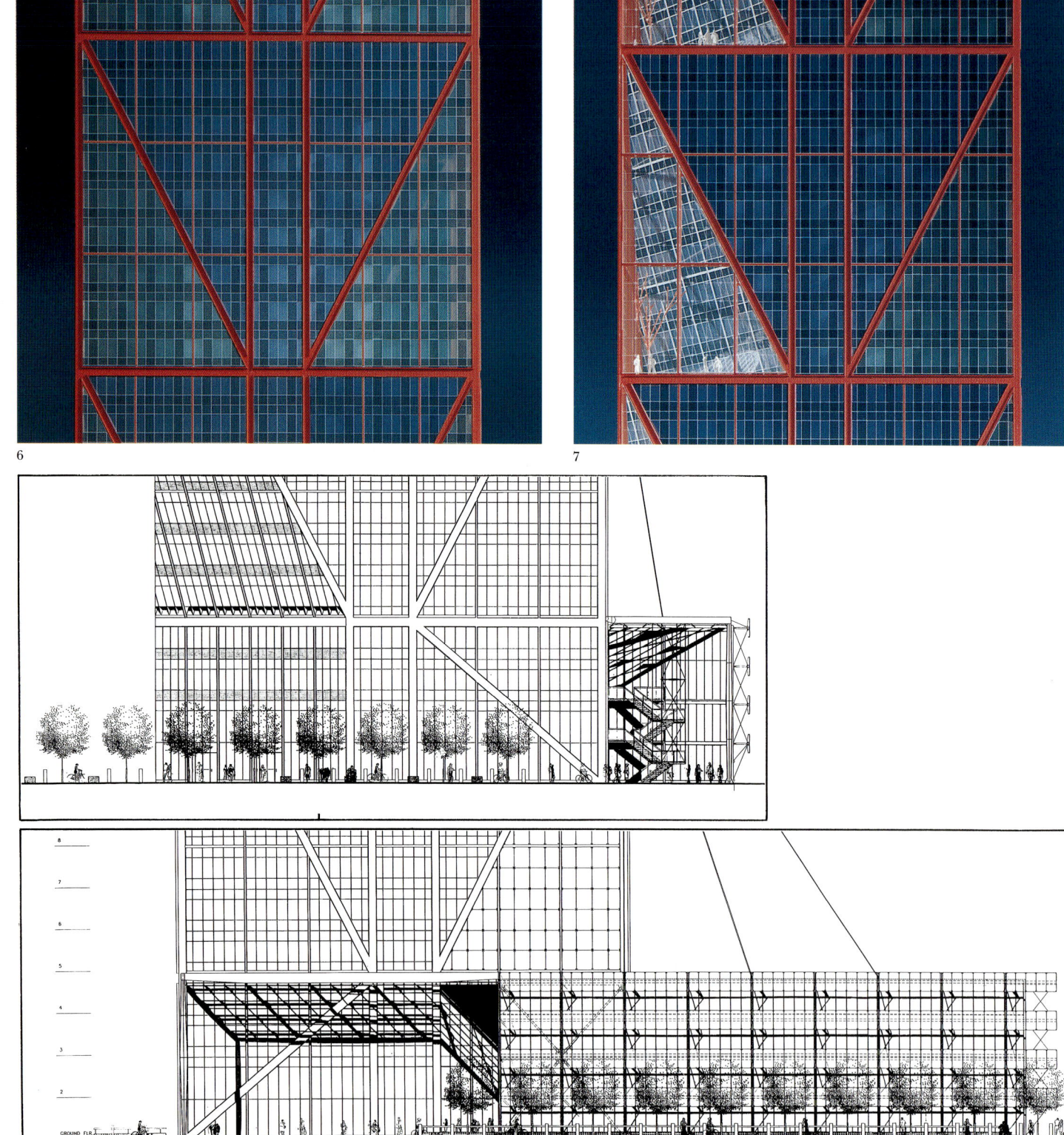

Endless Towers

Design 1995
Theoretical high-rise project
1,865,662 square feet
Mega-structure framing of steel box sections forming an external tube
Skin of reflective glass, patterned spandrel glass, high-performance clear glass at skygardens, solar glass at interior walls at skygardens

无穷尽发展的塔楼

设计　　1995
理论性研究的高层建筑项目
1,865,662 平方英尺
用钢箱形截面构成的巨型结构形成的一个外部筒体
反光玻璃幕墙，带图案的层间玻璃幕墙，空中花园处为高性能透明玻璃，空中花园处的内墙采用日光玻璃

This project consists of two similar towers in which the container is juxtaposed with the contained. The bold simplicity of the forms is inspired by the artist Brancusi.

Tower 1 expresses the structural endless column system in a bold graphic pattern. Tower 2 extends the immateriality of the container beyond the building's top into a roof garden.

The alternating symmetry of the chamfered corners defines a trussed tube. The mega-structure utilizes this architectural form to obtain efficient vertical and lateral load-resisting systems.

The skygardens provide quality space, panoramic views, and a more pleasant working environment. They are also an important part of the towers' environmental control systems, providing a buffer zone beteween the office space and the outdoor conditions.

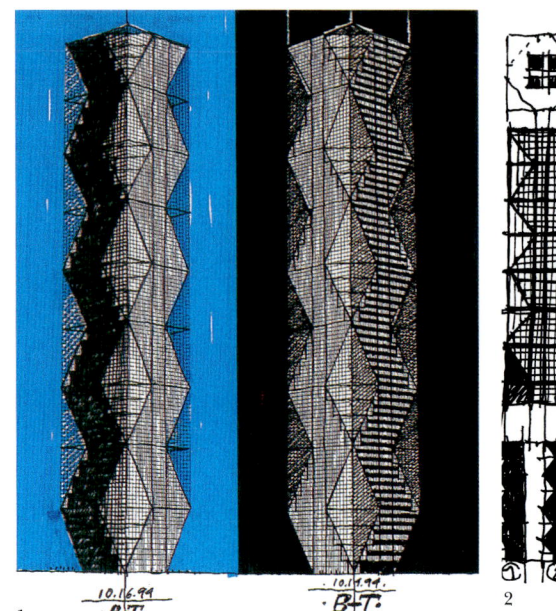
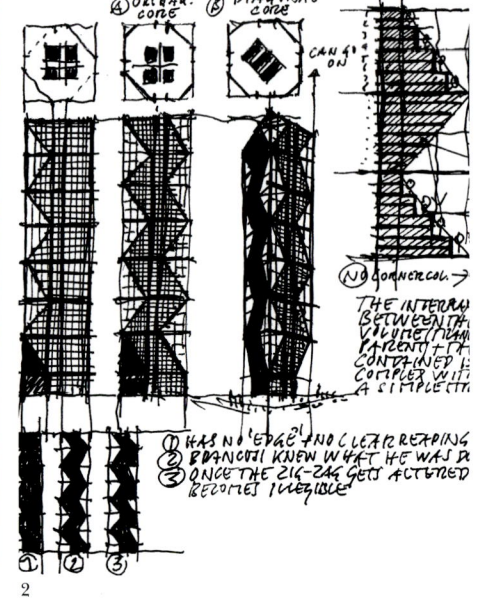

　　该项目由两座相似的塔楼所组成，在塔楼之中，容器与被包容物并置在一起。大胆而简洁的形式受到了艺术家布兰库斯的启发。

　　第一栋塔楼以一种鲜明的图形模式表达了在结构上可以无穷尽发展支柱体系。第二栋塔楼将容器的非物质性特征一直延伸超过建筑的顶部，形成了一个屋顶花园。

　　斜切的转角处形成的交替对称处理形成了一个桁架筒体。巨型结构利用这种建筑形式而形成了有效的垂直和水平抗压体系。

　　在空中花园形成了高品质的空间、全景式的景观以及一个更为愉悦的工作环境。它们也是塔楼的环境控制系统中的一个重要部分，在办公空间和户外环境之间形成了一个缓冲区。

1-2　构思草图
3　　在城市环境中建筑塔楼之间的相互影响

1-2　Concept sketches
3　　Interplay of building towers in the city context

Tokyu-Shibuya Tower

Design 1992
Tokyo, Japan
Tokyu Corporation
Associate Architect: Kume Sekkei Co. Ltd/Media Five
1,022,571 square feet
Structural steel, glass curtain wall
Granite floor and plaza

东丘－涩谷塔楼

设计　　1992
日本，东京市
东丘株式会社
合作建筑设计单位：久米摄敬有限公司/媒体五工作室
1,022,571平方英尺
钢结构，玻璃幕墙
花岗石楼面和广场

The slightly skewed plan of the Tokyu-Shibuya Tower generates an elegant, distinctive and omni-directional shape. The building form is accentuated further through inflections on the facade, and the top is sloped to complete the sculptural composition.

The building comprises two main components: the hotel/office tower and the wedding hall/conference center. A glass-covered porte-cochere separates the two components. The first floor of the conference center contains the lobby and the wedding hall. A main ballroom is on the second floor. An enclosed bridge over the porte-cochere at the mezzanine level connects to the hotel tower.

The porte-cochere also serves as the entry to the hotel. Restaurants and shops are at the ground floor, and a tea lounge is on the mezzanine level. A landscaped garden serves as a backdrop to this space. Express elevators take hotel guests from the ground floor to the main hotel lobby and front desk at the eighteenth floor.

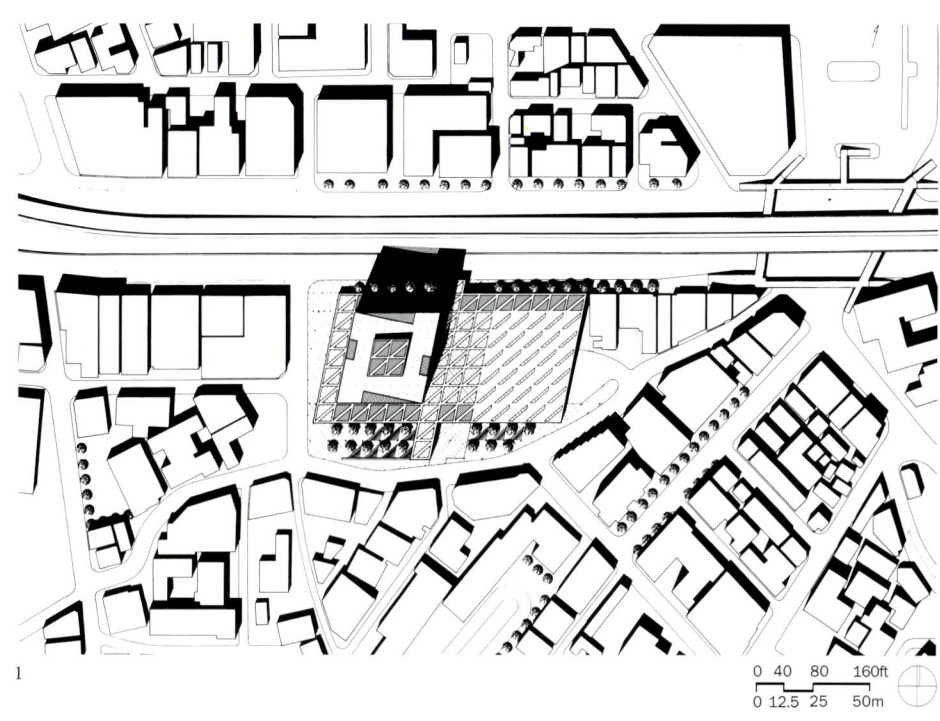

东丘－涩谷塔楼略微倾斜的平面产生了一个优雅的、鲜明的、全方向性的形体。这种建筑形式通过立面上的变形得到进一步的强化，并且建筑顶部也进行倾斜处理，以此来结束整个建筑的雕刻式的构图。

这栋建筑由两个主要部分组成：酒店/办公塔楼和婚宴大厅/会议中心。一个用玻璃顶覆盖的车辆通道将两个部分隔开。会议中心的一层设有大堂和婚宴大厅。一个大型的舞厅位于第二层上。在夹层标高上设有一个位于车辆通道上方的封闭式天桥，将会议中心与酒店塔楼连接起来。

车辆通道也被作为酒店的入口。底层设有餐厅和商场，在夹层上设有一个休闲茶室。一个景观花园被用作该空间的背景。高速电梯将酒店客人从底层送到位于第八层上的酒店大堂和前台。

1　总平面图
2　酒店/办公塔楼和会议中心表现图

1　Site plan
2　Expression of hotel/office tower and conference center

Hitachi Tower/Caltex House

Design/Completion 1988/1993
Savu Properties Pte.
Associate Architect: Architects 61
Hitachi Tower: 375,000 square feet
Caltex House: 400,000 square feet
Tower: steel frame and reinforced concrete core
Arcade: exposed steel frame
Aluminum panels, glass infill, structure glazing and marble floors

Located at the historic Change Alley Arcade, Hitachi Tower and Caltex House have a significance that must work on two levels: continuing an evolving urban context and establishing an identity on the Singapore skyline. Change Alley has historically been an important venue for changing money and purchasing local handicrafts. This arcade now finds itself as part of Singapore's ever-growing financial district.

The design of the towers creates a gateway to an important pedestrian street linking the sea and Collyer Quay with Raffles Square, the financial district and the towers that comprise the new Singapore.

The elements of Hitachi Tower—shaft and podium—reinforce the urban condition along Collyer Quay. The 38-story tower is rendered as a one-story frame with a setback at the twenty-fourth floor. A metal bay is juxtaposed with the background frame in order to introduce a scale element and reinforce the orientation to the sea.

Continued

日立塔楼/卡尔泰克斯大楼

设计／竣工　1988/1993
新加坡
佐武房地产开发公司
合作建筑设计单位：61建筑师事务所
日立塔楼：375,000平方英尺
卡尔泰克斯大楼：400,000平方英尺

日立塔楼和卡尔泰克斯大楼位于有着历史意义的变革巷廊道上，它们具有一个必须在两个层面上产生作用的重要性：延续一个不断进化的城市文脉以及在新加坡的天际轮廓线上体现其个性。变革巷在历史上是一个兑换货币和购买当地工艺品的重要场所。这个廊道现在已经成为新加坡持续发展的金融区中的一个部分。

该塔楼在设计上形成了通往一条重要的步行街的门户，这条步行街将大海和科利尔码头与构成新的新加坡市的莱佛士广场、金融区以及几栋塔楼连接了起来。日立塔楼上的要素（塔身和裙房）强化了沿着科利尔码头的城市环境。这栋38层塔楼被处理成一个单层的框架，在第24层处有一个向后退台的处理。一个金属排架与背景框架并列，以引入一种尺度元素以强化向着大海的朝向。

（待续）

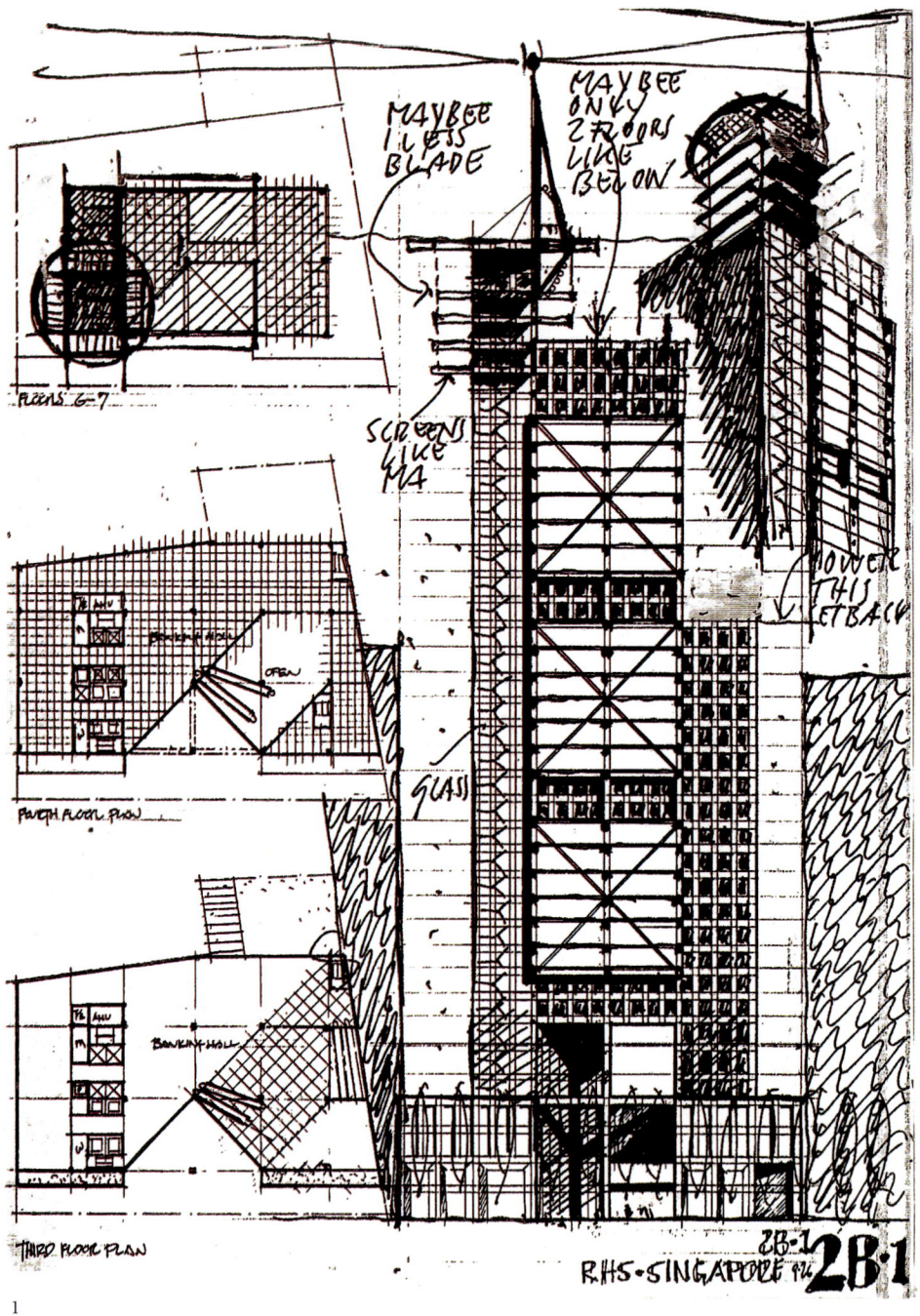

1　Hitachi Tower: concept sketch
2　Hitachi Tower: front/harbor elevation

1　日立塔楼：构思草图
2　日立塔楼：正立面／沿海港立面

2

Hitachi Tower/Caltex House 57

The "free" geometry of the ground floor reinforces pedestrian movement in and around the site. Small shops line the arcade to foster movement to Raffles Square. A cool palate of white, grey and blue is utilized as a welcome relief to the tropical atmosphere of Singapore.

The organization of Caltex House is that of a four-story retail podium and arcade with an office tower of 29 stories rising above. The rounded end of the tower offers views of Raffles Square, while also generating the top of the tower and providing a four-story entry portico.

Four levels of retail extend the retail component of Hitachi Tower in a style that is sympathetic in detail and material. The 72-foot high arcade space is topped with a glass roof offering protection from the elements, but providing ample natural light. The enclosure of the tower is rendered in metal and glass, similar to Hitachi Tower.

底层的"自由"几何构图加强了行人在场地内及周围的活动。沿着小商店廊道一字排开，引导着通往莱佛士广场的人流活动。建筑上采用了一种由白色、灰色和蓝色构成的冷色调，很好地缓解了新加坡的热带气候，颇受人们的欢迎。

卡尔泰克斯大楼由一个四层楼的商业零售裙房和廊道以及位于其上方的29层高的办公塔楼所构成。塔楼的圆形外墙处理让人们能够看到莱佛士广场，同时也产生了塔楼的顶部造型，并形成了一个4层高的入口门廊。

4层的商业零售空间采用了一种在细部和材料上相一致的方式来延伸日立塔楼中的商业零售部分。72英尺高的廊道空间上方采用了一个玻璃顶，它既起着对空间的保护作用，同时也提供了充足的自然采光。塔楼的外墙部分采用了金属和玻璃材料来表现，与日立塔楼有着类似的效果。

3

3 Hitachi Tower: curtain wall detail
4 Hitachi Tower: harbor view from the south-east with metal panel service wall
5 Hitachi Tower: harbor view from the north-east

3 日立塔楼：幕墙细部
4 日立塔楼：从东南面看港口和金属面板设备墙
5 日立塔楼：从东北面看港口

4

5

Hitachi Tower/Caltex House   59

6

7

8

6 Level 2 galleria toward Hitachi
7 Entrance Caltex/Subway
8 Caltex glass wall entrance
9 Hitachi Tower: elevator cab
10 Hitachi Tower: elevator lobby
11 Hitachi Tower lobby

6 通往日立塔楼的二层走廊
7 卡尔泰克斯/地铁入口
8 卡尔泰克斯大楼的玻璃墙入口
9 日立塔楼：电梯轿厢内部
10 日立塔楼：电梯间
11 日立塔楼大厅

9

10

11

12 Retail galleria section
13 Retail passageway from Caltex House
14 Level 3 galleria toward Caltex House
15 Caltex House: concept sketch
16 Caltex House: skylobby

12 商业零售空间剖面图
13 从卡尔泰克斯大楼方向过来的商业零售区通道
14 通往卡尔泰克斯大楼的第三层商业零售空间
15 卡尔泰克斯大楼：构思草图
16 卡尔泰克斯大楼：空中大堂

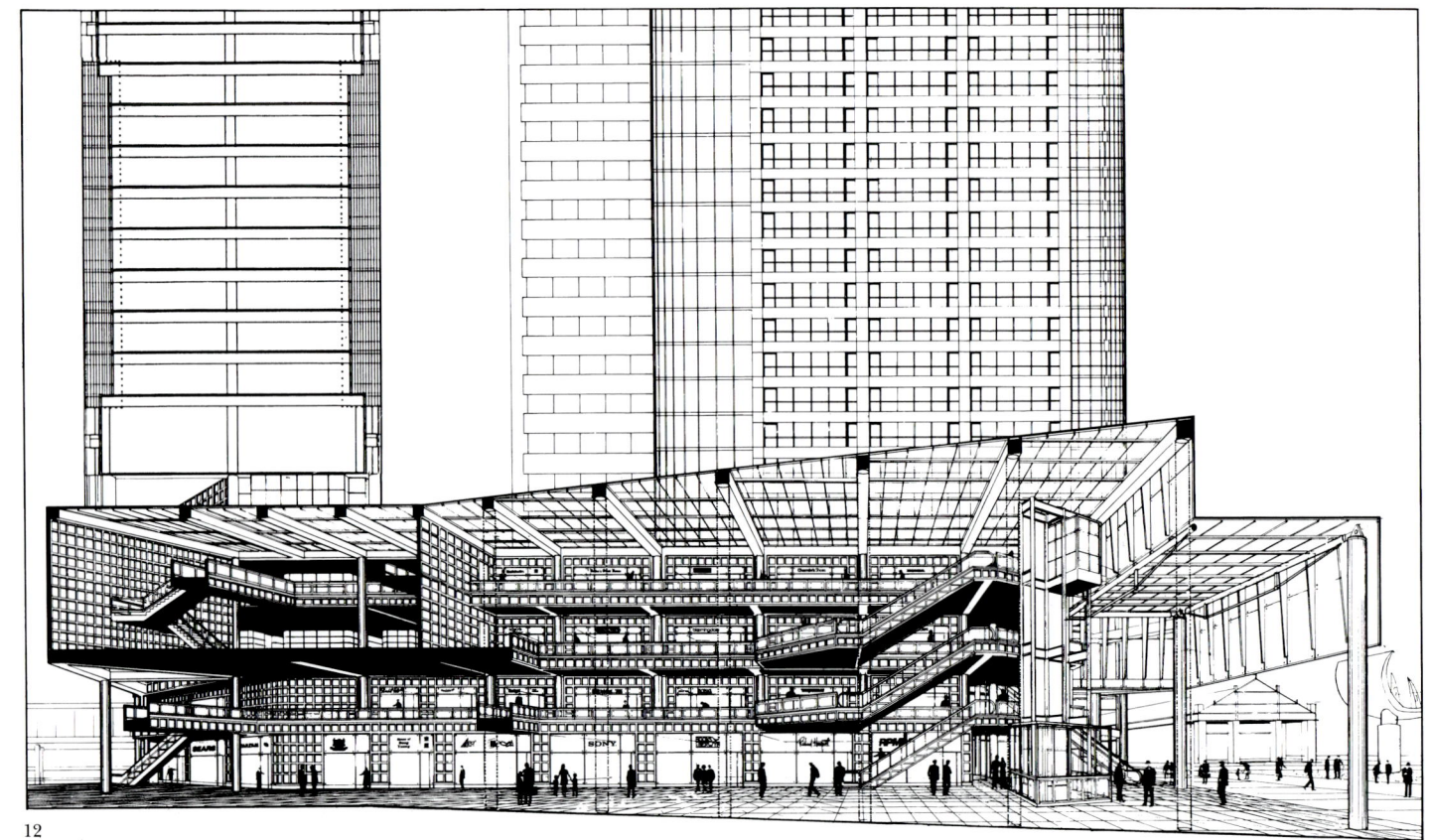

12

13

14

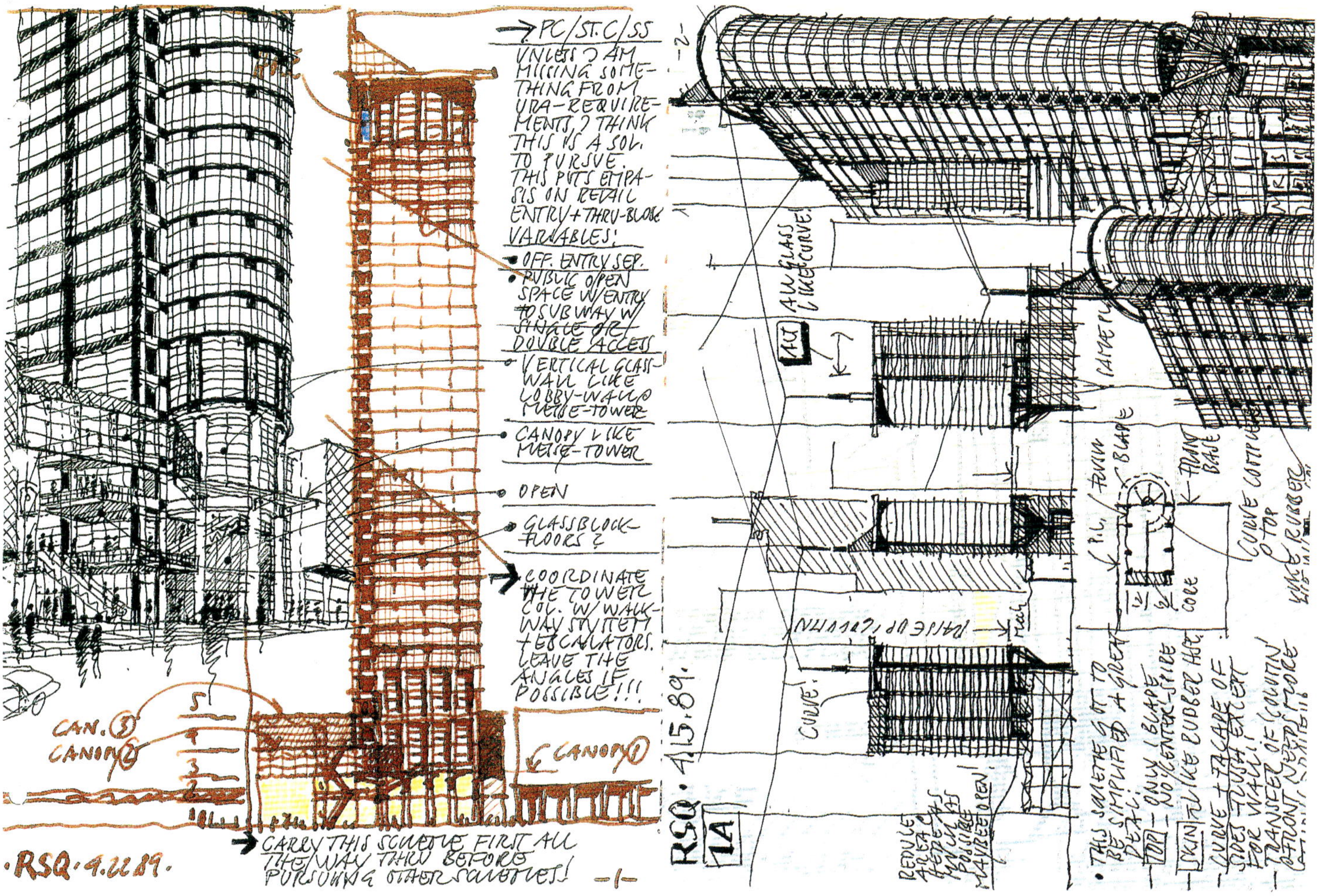

Hitachi Tower/Caltex House 63

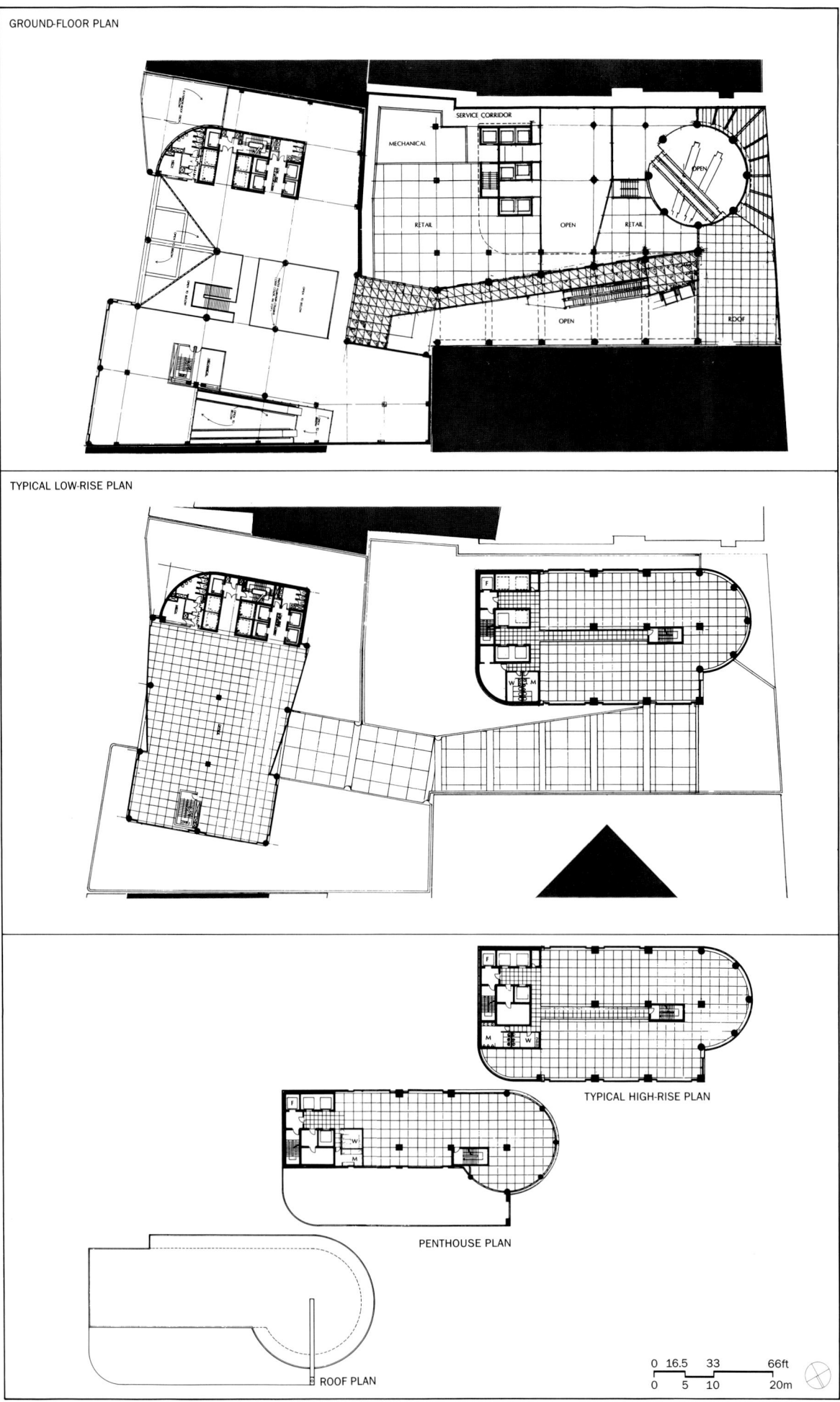

17　Tower plans
18　Caltex House from the north
19　Caltex House: brow project spandrel
20　Caltex House: glass/metal banding

17　塔楼平面图
18　从北面看卡尔泰克斯大楼
19　卡尔泰克斯大楼：顶端突出的檐口拱肩
20　卡尔泰克斯大楼：玻璃/金属水平带

19

20

18

Fort Canning Tower

Design 1992
Singapore
Singapore Telecom
Associate Architect: Architects 61
Height: 950 feet
Steel frame with stainless steel cladding

堪宁堡塔楼

设计　　　1992
新加坡
新加坡电信公司
合作建筑设计单位：61 建筑师事务所
建筑高度：950 英尺
钢框架结构，不锈钢面层处理

The tower is designed to be adaptable to meeting changing social, political and technical needs. Formally the tower is a helix-wound perforated steel plate coupled to the mast with outriggers. The negative surface is comprised of a lattice steel structure, making the tower at once solid and transparent. The steel tower rests upon a granite plinth housing radio equipment and the visitors' lobby.

The tower is sited at the Terrace View Garden which serves as the forecourt. A rolling hedge garden completes the circular form, and the tower is sited asymmetrically within this composition.

The structure embodies the principle of tube within tube to provide a stiff, slender structural platform to satisfy the deflection limitations of the telecommunications systems. These systems are housed in discrete pods at the most efficient level in terms of structural considerations and distance from antennas.

　　该塔楼在设计上是为了满足不断变化的社会、政治和技术上的需要。在建筑形式上，塔楼是一个螺旋形环绕上升的带孔钢板，通过悬臂挑梁与中心立柱相连。塔楼的内层墙面由格子状的钢结构组成，使塔楼显得既坚固又通透。这个钢制的塔楼竖立在一个花岗石的基座上，基座里设置了无线电设备用房和供参观者使用的大堂。

　　该塔楼坐落于台地观景花园中，该花园成为塔楼的前庭。一个由起伏的树篱围合起来的花园完善了其圆形，而塔楼则不对称地位于该构图之中。

　　该建筑的结构体现了筒中筒的原则，从而形成了一个坚固而细长的结构平台来满足通信系统对偏移限制方面的要求。这些通信系统被布置在分隔开来的舱室中，这些房间根据对结构的考虑和与天线之间的距离要求而设置在最有效的标高上。

1

1　通信塔楼成为新加坡的一个标志

1　The communications tower becomes a symbol for Singapore

IHZ Tower

Design 1992
Düsseldorf, Germany
Premier GmbH Bouygues
880,100 square feet
Steel with major and minor frame development
Glass and aluminum curtain wall, aluminum columns, granite, textured metal panels and aluminum grid structures

IHZ 塔楼

设计　　1992
德国杜塞尔多夫市
布伊格斯第一股份有限公司
880,100平方英尺
钢结构及大型和小型框架
玻璃和铝材幕墙、铝材支柱、花岗石、有纹路的金属面板和铝制栅格结构

The site for IHZ Tower lies in an evolving master-planned development in Düsseldorf. The primary urban planning goal is not to conform to the existing context, but to lead and invigorate the realization of the future-oriented project.

The form of IHZ Tower derives from the functional need to provide clear, efficient, well-lit office space. Composed of two parallel office wings connected by a core and service areas, the tower features floor plates offering well-oriented office layouts, complete flexibility in leasing, and economy in construction. At lower levels the tower is connected to an auxiliary structure providing for single-floor tenants.

Terraces created at levels 14 and 26 provide feature areas for the restaurant, cafeteria, and club facilities, which are located to take advantage of the exterior space.

The auxiliary building contains a business center and other communal functions. A central lobby/atrium separates the two project components.

IHZ塔楼所在的基地位于杜塞尔多夫市的一个正按照总体规划实施的开发区中。这一城市规划的主要目标不是要与现有的文脉保持一致，而是要引导和鼓励那些面向未来的项目的实现。

IHZ塔楼的造型是来自于功能上的需要，即提供清晰、高效和采光良好的办公空间。该塔楼由两个平行的办公楼侧翼组成，通过一个核心和设备区域将两个部分连接起来。塔楼的楼板形成了具有良好朝向的办公室布局，在租赁方面也有着很好的灵活性和在建筑施工上的经济性。塔楼的下面部分与一个附属建筑物连在一起，该附属部分提供单房间给小租赁户使用。

在第十四层和第二十六层上的平台形成了一些用作餐厅、自助餐厅和夜总会等设施的有特色的地方，这些设施的位置都充分利用了外部空间。

附属建筑物中设置了一个商务中心和其他公共性功能。一个中央大堂/中庭将这两个建筑部分区分开来。

1

1 从东北方向看建筑，展示出结构和外表面之间的相互影响
2 构思草图

1 View from the north-east showing the interplay of structure and skin
2 Concept sketch

2

120 North LaSalle

Design/Completion 1988/1991
Chicago, Illinois
Ahmanson Commercial Development Corporation
576,900 square feet
Placed concrete, steel at vaults
Polished Rockville, honed Raphaela, and flamed Galizia granites; gray tinted glass and patterned fritted spandrel glass

120 北拉萨尔大楼

设计/竣工　1988/1991
伊利诺伊州，芝加哥市
阿曼森商业开发有限公司
576,900平方英尺
现浇混凝土结构，拱顶部分为钢结构
抛光罗克维尔花岗石、细磨拉斐拉花岗石、火烧加利齐亚花岗石
灰色玻璃和压花热熔窗间玻璃

The building's location on a narrow mid-block site in LaSalle Street, one of the city's most highly defined urban spaces, generated its particular forms and spaces.

The inherent asymmetry of the site led to a structure that is asymmetrical in plan and elevation. All fixed elements are contained in a compact side-core to provide adjacent unobstructed office space of maximum flexibility. A curved bay extends over LaSalle Street, responding to the limited corner condition at Court Place. The projecting bay affords excellent views down LaSalle Street and to the river, giving the space a unique character.

A coved loggia with a large mosaic mural brings the public realm into the site. A projecting trellis extends through Court Place and to the entrances at LaSalle and Wells Streets, enhancing the pedestrian experience of this well-traveled mid-block link to City Hall.

该建筑位于拉萨尔大街的上一个狭窄的街区中央的地块上，该街道是这个城市中最为突出的城市空间，从而形成了该建筑独特的形式和空间。

该建筑基地所固有的不对称性决定了建筑物在平面和立面上的不对称。所有的固定元素都被包含在一个紧凑的边核心区之内，从而形成了连在一起的无所阻挡并具有最大的灵活度的办公空间。一个曲面的凸窗延伸在拉萨尔大街立面上，与法院广场上的转角处理相互呼应。悬挑出来的凸窗能够提供俯视拉萨尔街和眺望河流的美景，并赋予了这个空间一个独特的特征。

一个有着一幅大型马赛克壁画的内凹式凉廊将公众领域引入到建筑场地之中。一个悬挑的格架从法院广场上延伸过来，一直到拉萨尔街和韦尔斯街的入口处，从而加强了行人对这个熙熙攘攘的与市政厅相连的街区中段上的步行的感觉。

1–2　构思草图
3　从人行道上看曲面的挑廊和凸窗

1–2　Concept sketches
3　Pedestrian view of curved loggia and projecting bay

4 Gridded north wall, trellis, projecting glass bay and curved blades at top
5 Asymmetrical stone service core
6 Entrance lobby
7 Elevator lobby
8 Streetscape

4 网格状的北向外墙、格架、凸出的玻璃窗以及顶部弯曲的片墙
5 不对称的石质设备筒体
6 入口大厅
7 电梯厅
8 街道景观

5

6

7

8

Messe Tower/Messe Hall

Design/Completion 1985/1991 (Tower), 1985/1988 (Exhibition Hall)
Tower: TishmanSpeyer Properties of Germany, LP
Exhibition Hall: Messe Halle Messe Frankfurt GmbH
Tower: 971,000 square feet
Exhibition Hall: 820,680 square feet
Tower: Poured-in-place concrete
Panelized curtain wall with red granite, painted aluminum mullions, patterned and clear glass
Exhibition Hall: Poured-in-place concrete, steel trusses
Metal and glass panel wall systems

博览会塔楼/博览会大厅

设计/竣工　　1985/1991（塔楼），1985/1988（展厅）
塔楼：德国提史曼史拜尔房地产公司
展览大厅：法兰克福世博会股份有限公司
塔楼：971,000 平方英尺
展厅：820,680 平方英尺
塔楼：现浇混凝土
红色花岗石镶饰的幕墙，上漆铝材竖框
压花和透明玻璃
展览大厅：现浇混凝土，钢结构桁架
金属及玻璃板墙体系统

The client's brief called for a new exhibition hall with 215,000 square feet of exhibition space, an entry building, a rentable office building, and 900 parking spaces. The 1909 Festhalle built by Friedrich von Thiersch, and the post-war Kongresshalle are immediately adjacent. Together with these two buildings, the new Exhibition Hall (Halle 1) forms a border to the outdoor exhibition space, which is open to the avenue and park of the Friedrich-Ebert-Anlage.

The office building is freestanding and is placed like a campanile among the composition of low buildings. The tower is raised high above the ground to create a symbolic gate to the Messe.

The architectural themes—long span hall, arcade, pavilion, tower, campanile, and gate—are elaborated in the forms of the buildings and their subtle relationships. Pure geometric forms—the square, circle, cylinder, and pyramid—are interrelated, rotated, transformed and elaborated in the materials and detailing.

建设方提供的设计任务书要求建设一个拥有 215,000 平方英尺展览空间的新展厅、一栋入口建筑、一个可供出租的办公大楼，以及有 900 个车位的停车场。由弗雷德里希·冯·希尔什在 1909 年设计的典礼大厅和战后的国会大厦与该建筑非常接近。新展厅（一号展厅）与这两栋建筑共同构成一个室外展览空间的边界，并朝着弗雷德里希公园区的街道和公园方向开放。

办公大楼为独立式的，宛如一座钟塔耸立在一片低矮的建筑群中。该塔楼高出地平面许多，形成了博览会的一个象征性的入口。

建筑主题（大跨度的大厅、廊道、亭阁、塔楼、钟塔及大门等）通过建筑的形式和其微妙的关系被精心地表达出来。纯净的几何形式——方形、圆形、圆柱体及方锥体等彼此相互联系，运用了旋转、变形等手法，并通过材料和细部处理来予以精心表现。

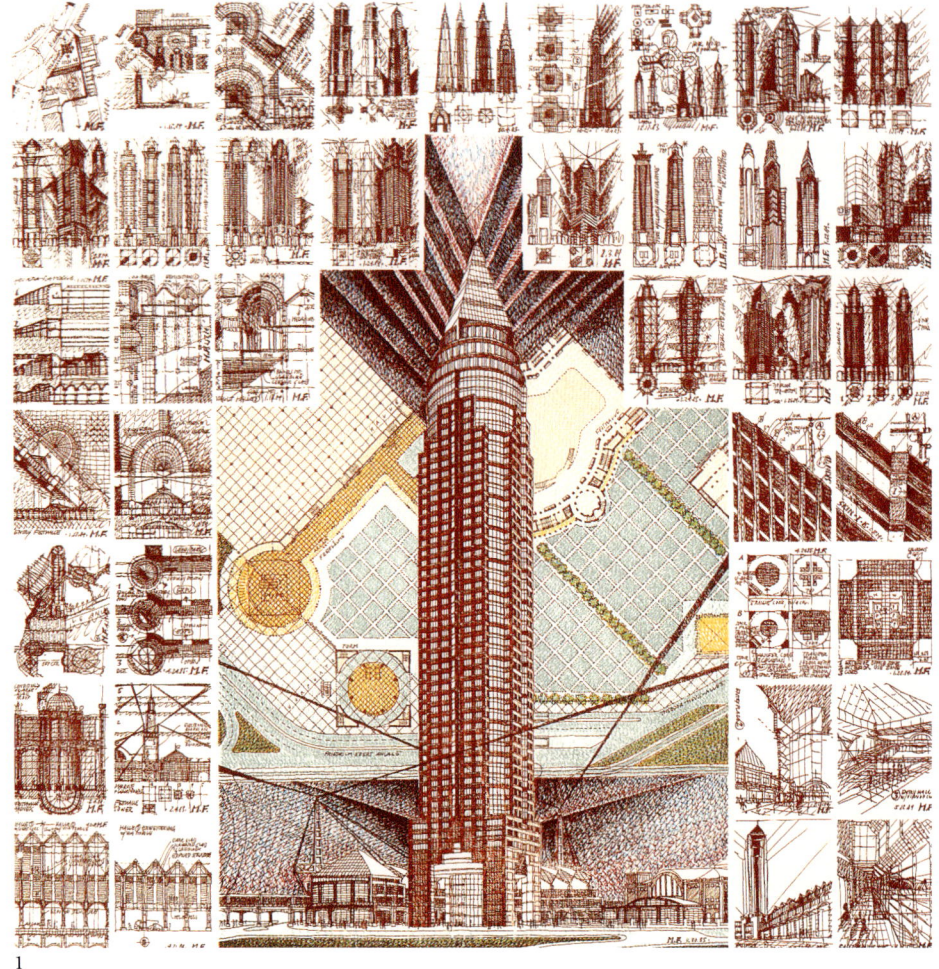

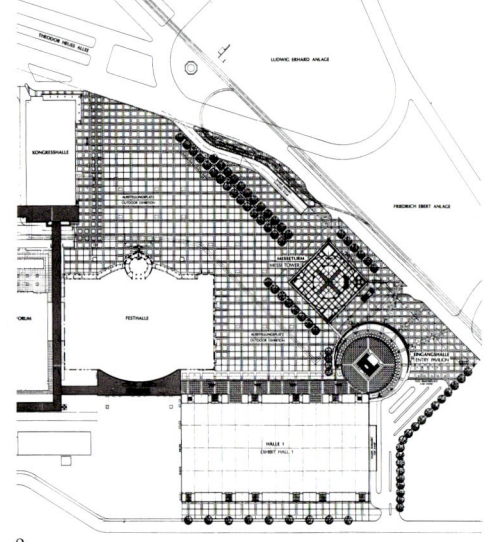

1　塔楼草图
2　总平面图
3　法兰克福附近的博览会交易市场

1　Tower sketches
2　Site plan
3　Messe fairground relative to the city of Frankfurt

4	Entrance gate	4	入口大门
5	Cylindrical glass entrance with canopy	5	带雨篷的圆柱体玻璃入口
6	Lobby	6	大厅
7	Pyramid top with glass cylinder	7	玻璃圆柱体上的金字塔式顶部
8	Public plaza	8	公共广场
9	Section/plans	9	剖面图/平面图

7

8

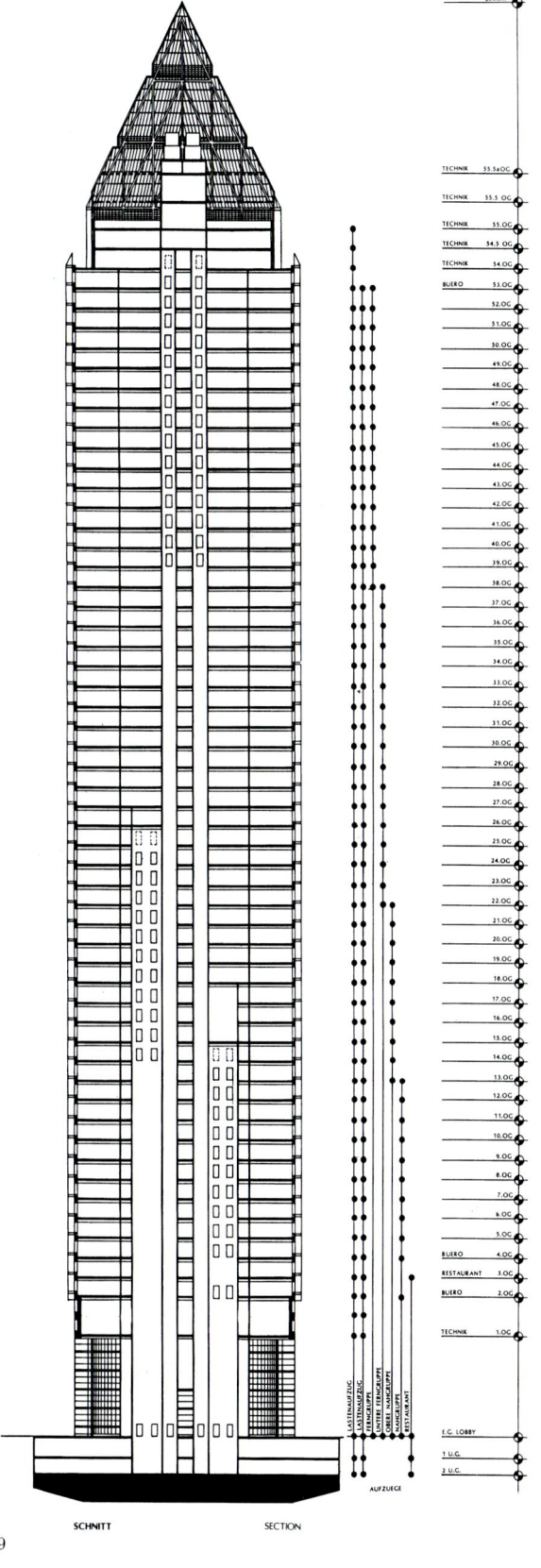

9

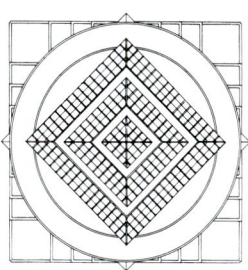

ROOF PLAN

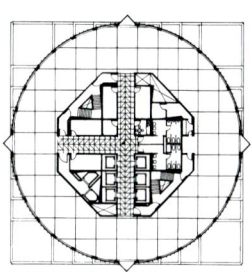

HIGH-RISE CYLINDER

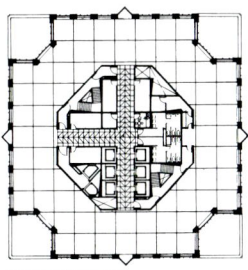

HIGH-RISE PLAN

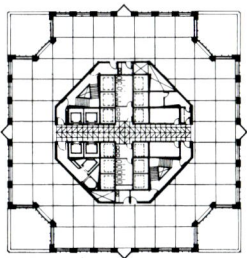

LOW-RISE PLAN

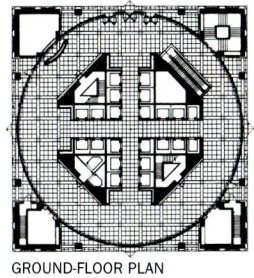

GROUND-FLOOR PLAN

Messe Tower/Messe Hall 75

10 City view with registration pavilion showing punched stone wall with glass cylinder
11 Exhibition Hall and registration pavilion
12 Aerial view of Exhibition Hall

10 城市景观，近处是报到登记亭，底部是穿孔石墙体，上面为玻璃圆柱体
11 展览大厅及报到登记亭
12 展览大厅的鸟瞰

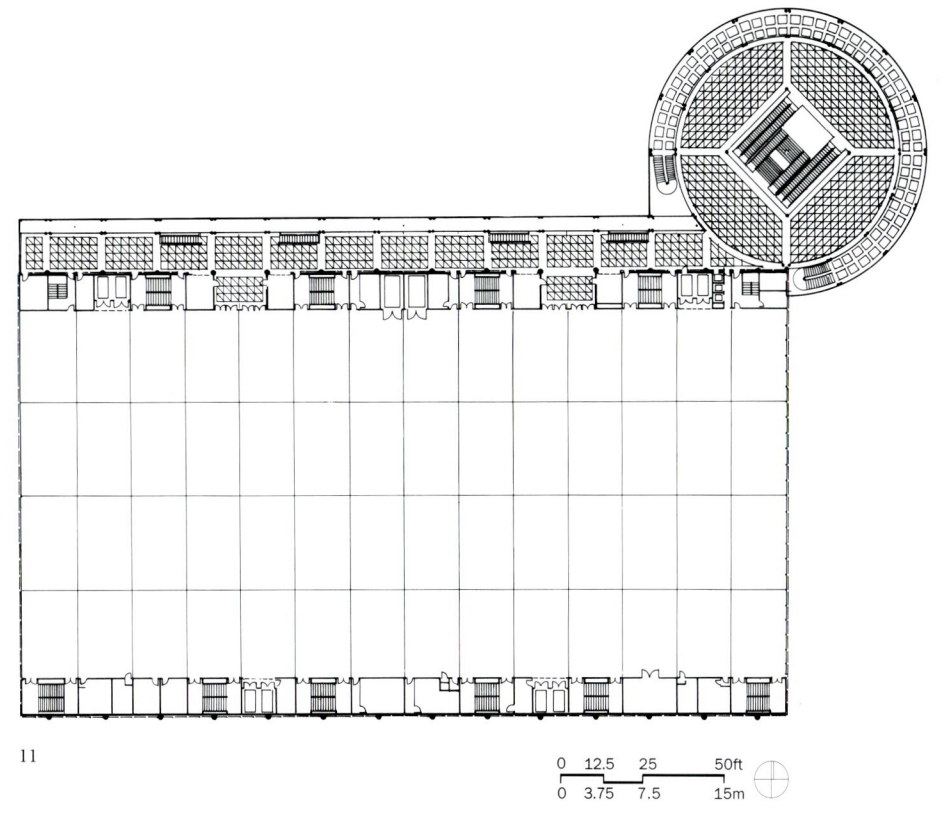

11

12

Messe Tower/Messe Hall 77

13

14

15

13 Registration pavilion and public plaza	13 报到登记亭及公共广场
14 Exhibition Hall truss/column connection	14 展览大厅的桁架与柱子的连接方式
15 Exhibition Hall elevation	15 展览大厅的立面
16 Truss/column connection detail	16 桁架与柱子连接处的细部
17 Glass end wall	17 玻璃山墙
18 Registration/entry pavilion interior	18 报到登记亭/入口亭的内部
19 Exhibition Hall interior	19 展览大厅的内部
20 Exterior roof truss	20 室外的屋面桁架
21 Roof truss, mechanical, service corridor	21 屋面桁架及机械设备服务走廊

16

17

18

19

20

21

Fountain Square West

Design 1991
Cincinnati, Ohio
The Galbreath Company
1,200,000 square feet
Steel frame with concrete core
Steel superstructure clad in stainless steel; curtain wall infill of insulating vision glass, fritted spandrel glass; fluoropon-coated aluminum

喷泉广场西侧大楼

设计　1991
俄亥俄州，辛辛那提市
盖尔布里斯公司
1,200,000平方英尺
钢框架结构，混凝土核心
用不锈钢覆面的钢制上部结构，不透明窗间玻璃幕墙填充材料，热熔层间玻璃，含氟涂层铝构件

Fountain Square West is one of few remaining sites in downtown Cincinnati of great urban significance, occupying the length of Fifth Street with prominent frontages along Race and Vine Streets. This project imaginatively integrates all functions within a mixed-use complex. The high-rise tower contains 750,000 square feet of corporate office space and a 250-room hotel. Some 175,000 square feet of retail is provided on three levels within the base, with parking for 750 automobiles below.

The retail base concentrates its focus on Race Street, the traditional downtown retail street, while the office tower and hotel functions focus on Fountain Square itself. Fountain Square West would rise above the city, reshaping the downtown skyline, and becoming the focal point for Cincinnati.

喷泉广场西侧大楼是在辛辛那提市区保留下来的为数极少的几个具有重大城市意义的地方之一，它占据了整条的第五大街，其引人注目的建筑立面顺着瑞斯大街和怀恩大街展开。该项目巧妙地将所有的功能都整合在一个多功能综合建筑之中。高层塔楼中有750,000平方英尺的公司办公空间及一个250个房间的旅馆。在基座内的3层楼中有大约175,000平方英尺的商业零售区，其下为一个可停放750辆车的停车场。

用作商业零售区的基座部分将其重点放在瑞斯大街一侧，该大街是一个传统的市中心区商业零售街道，而办公塔楼及旅馆则集中在喷泉广场之上。喷泉广场西侧大楼将在城市上空高高耸立，重新构成市中心地段的轮廓线，也将成为辛辛那提市的焦点。

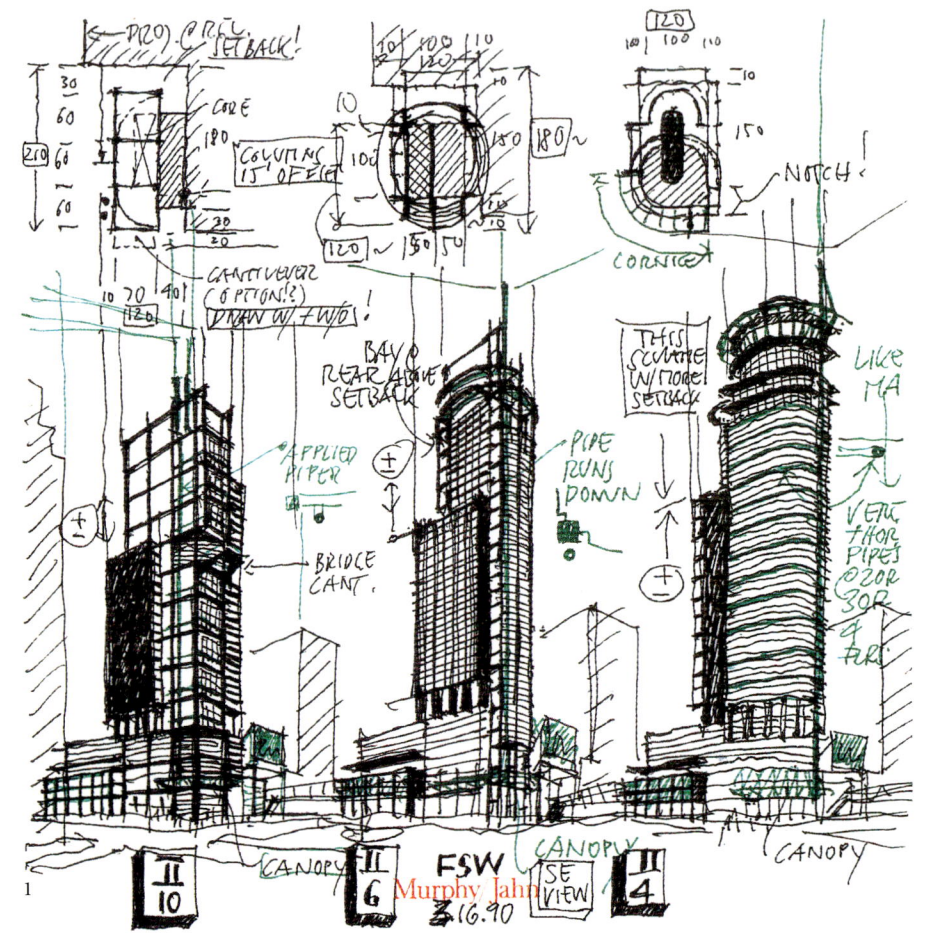

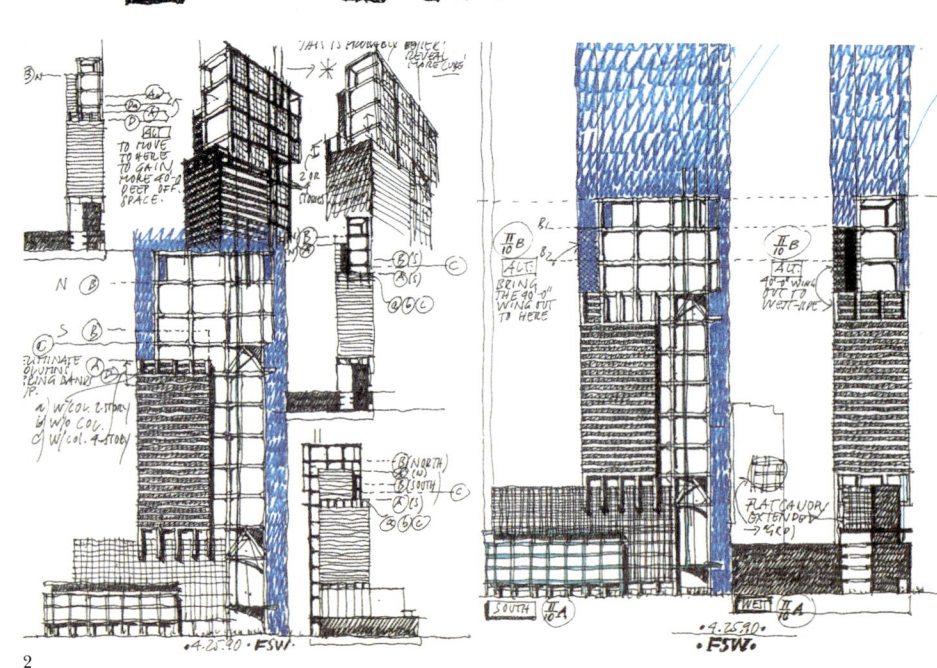

1 Concept sketch of building alternatives
2 Concept sketch of skin detail
3 City context: street view
4 Projecting front bay

1 表现多个建筑方向的建筑构思草图
2 外观细部的构思草图
3 城市环境：街道景观
4 挑出的开间部分

3

4

Fountain Square West 81

FAA O'Hare Control Tower

Design 1991
Chicago, Illinois
Chicago O'Hare International Airport/Department of Aviation/City of Chicago
Concrete mast with steel outriggers at perimeter
Clear glass infill panels

联邦航空局奥黑尔控制塔楼

设计　1991
伊利诺伊州，芝加哥市
芝加哥奥黑尔国际机场／航空部／芝加哥市
混凝土网架结构，周围设置了外挑钢制支架
透明玻璃填充板

The new control tower is the landmark of O'Hare International Airport, providing a highly visible point of orientation in the airport landscape.

The core of the tower is a concrete column which contains the elevator, cables, and technical equipment, with the staircase wrapped around it. The surrounding steel support structure is connected to the core by horizontal posts, and stayed to the foundations. A glass curtain covers the steel elements, creating a shape that reflects the functions of the tower.

During the day, the shape of the tower is dictated by the glass surfaces, transparency and reflection creating an interesting dialogue. At night, artificial lighting creates depth and transforms the tower into a luminous object—a beacon for O'Hare.

　　新的控制塔是奥黑尔国际机场的地标性建筑物，它在机场地貌之中形成了一个能被清楚地看到的定位点。

　　塔楼的中心是一个混凝土圆柱，其中容纳了电梯、电缆及技术设备，柱子外侧由楼梯间包裹起来。其四周的钢制支撑结构通过水平杆与塔楼核心部分相连，并由基础来支撑上部结构。采用一层玻璃幕将钢构件遮盖起来，最终形成一种可以反映出该塔楼的功能的外部造型。

　　在白天，玻璃表面确定了塔楼的外形，其透明性和反射性形成一种有趣的对话。在夜里，人工照明形成一种深度感，将塔楼变成一个发光物体——那就是奥黑尔机场的灯塔。

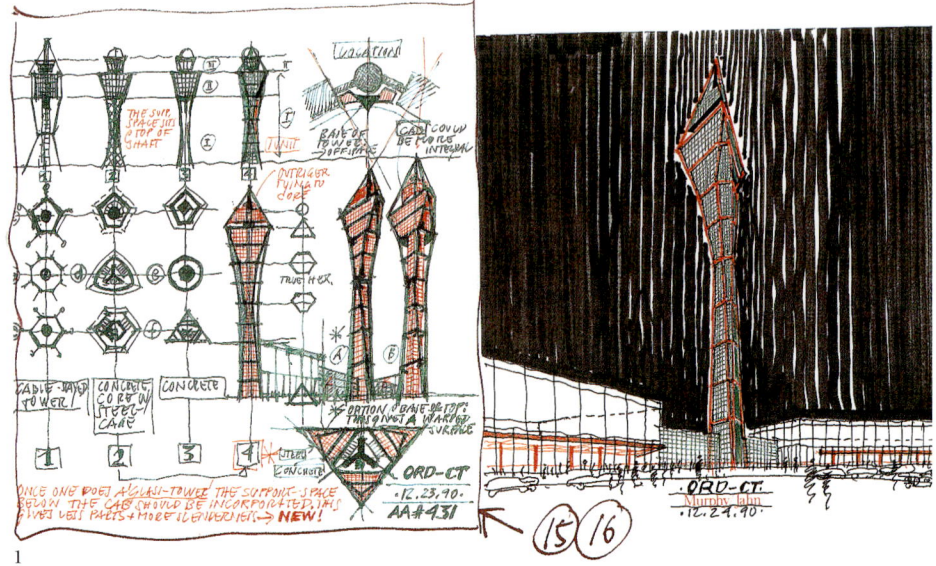

1

2

1　构思草图
2　从道路方向看塔楼

1　Concept sketches
2　View from roadway

Kuala Lumpur City Center 1

Design 1991
Kuala Lumpur, Malaysia
Seri Kuda Sdn. Bhd.
8,000,000 square feet
Structural steel, glass and metal curtain wall

吉隆坡一号城市中心

设计　　1991
马来西亚，吉隆坡
赛里奎达·塞德巴德
8,000,000平方英尺
钢结构，玻璃与金属幕墙

The design for the Kuala Lumpur City Center, sited at the former Royal Racecourse, is an attempt to re-examine the planning of dense urban multi-use centers. The development is the initial phase of the Kuala Lumpur City Center Master Plan.

The plan includes three office towers, a retail block and a hotel organized along a wide "main street", or spine, identified by its undulating roof structure. The retail is developed as a series of arcades and town squares emerging from and connecting to the main street.

The 95-story Petronus Tower and a smaller 50-story office slab front onto the "main street". A third office tower, 85 stories in height, is located along Jalan Ampang adjacent to the proposed Civic Center. The Convention Hotel is located to the south of the Petronus Tower. Each building has a distinct image which, through its technical features and use of color, establishes a language and builds a relationship with the larger urban context.

吉隆坡城市中心坐落于前皇家赛马场的位置上，该设计是重新检验高密度的城市多功能中心规划的一个尝试。此开发项目是吉隆坡城市中心总体规划的启动阶段。

该设计包括沿一条宽敞的"主街"或称主轴的三栋高层办公楼、一栋商业零售大楼，以及一个旅馆，以其起伏变化的屋面结构来形成该建筑群的特色。零售区被设计成从"主街"上延伸出来的并与其相连的一系列廊道和城市广场。

95层高的佩特罗努斯大楼和一座规模稍小的50层高的办公楼朝向"主街"布置。第三幢85层高的办公楼沿加兰·昂邦大街展开，与被提议的市政中心相邻。会议酒店位于佩特罗努斯大楼的南面。每幢建筑物均有着独特的形象，通过其技术特征及对色彩的运用来确立一种语言，并与更大范围的城市文脉之间建立起一种联系。

1　Interplay of individual tower element, hotel and retail

1　各独立的塔楼要素、酒店和零售区之间的相互影响

Barnett Center

Design/Completion 1988/1990
Jacksonville, Florida
Paragon Group Inc.
1,000,000 square feet
Cast-in-place reinforced concrete frame
Color-coated aluminum curtain wall system with reflective blue glass and Fjiord blue granite

巴尔内特中心

设计/竣工　　1988/1990
佛罗里达州，杰克逊维尔市
帕拉贡集团公司
1,000,000平方英尺
现浇钢筋混凝土框架结构
彩色涂层铝材幕墙，蓝色反射玻璃及费涅尔德蓝色花岗石

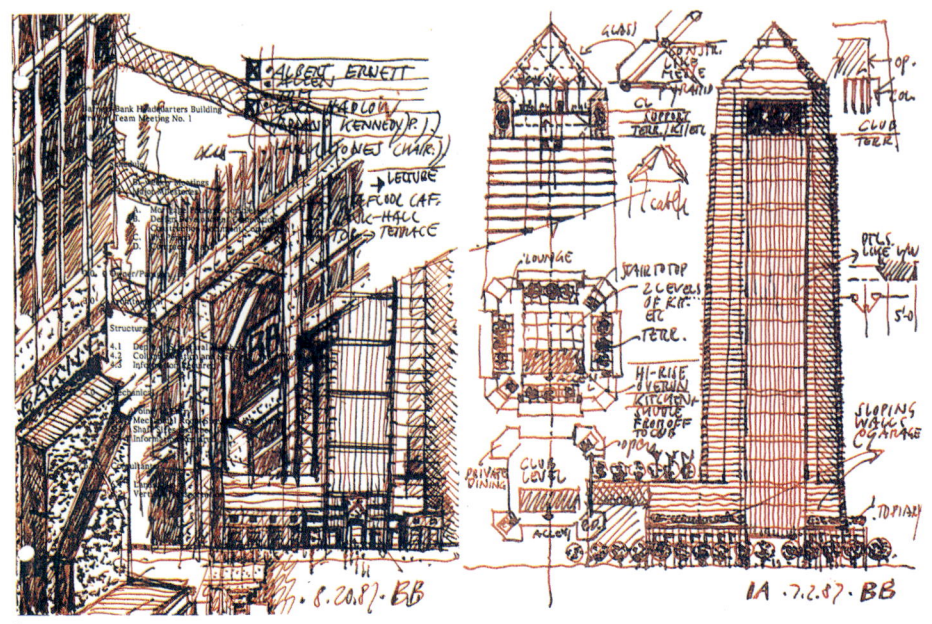

1

The brief for Barnett Center called for a soaring sculpture reaching a new height for Jacksonville, Florida. The building was designed as an obelisk, a rich traditional architectural form symbolizing achievement, power, glory, and victory. It is a meaningful symbol of the position of Barnett Bank within the financial community. Its elegant exterior treatment of horizontal and vertical patterns of blue-gray granite with silver mullions and blue tinted glass expresses the timelessness of an established financial institution.

The 42-story tower contains 710,000 square feet of office space, as well as parking for 654 cars, a spacious lobby, and a grand banking hall. The building is surrounded by a landscaped plaza giving access to ground floor retail and banking functions and the two-story lobby.

巴尔内特中心的设计任务书要求为佛罗里达州的杰克逊维尔市修建一座腾飞向上并达到一个新的高度的雕塑作品。该建筑被设计成一个方尖碑的形象，这是一种极具传统特征的建筑造型，它象征着成就、力量、光荣及胜利。它也成为对位于金融区的巴尔内特银行的地位的一种极富意义的象征物。该建筑在外墙上用蓝灰色花岗石配合银色竖框与蓝色玻璃形成的水平和垂直图案处理，表现出这一有着很高声誉的金融机构的永恒性特征。

42层高的塔楼中有710,000平方英尺的办公空间，还有可容纳654辆车的停车库、一个宽敞的入口门厅，以及一个华丽的银行业务大厅。大楼被一个有绿化处理的露天广场所环绕，人们可穿过广场到达一楼的零售区、银行业务区及两层高的门厅部分。

2

1 Concept sketches
2 Entrance
3 An obelisk on the Jacksonville skyline

1 构思草图
2 入口
3 在杰克逊维尔市的天际轮廓线中的方尖碑

Yokohama Waterfront MM21 Project

Design 1990
Yokohama, Japan
Mitsui & Co.
Associate Architect: AAS Associates International Co.
4,500,000 square feet
Tower portions steel moment connected frame base; steel moment connected frame
Painted aluminum panel and structure enclosure with clear and patterned fritted glass

横滨滨水地带 MM21 工程

设计　　1990
日本，横滨
三井公司
合作建筑设计单位：AAS 合伙人国际公司
4,500,000 平方英尺
塔楼部分：钢力矩连接框架裙房；钢力矩连接框架
涂层铝板及透明和压花热熔压花玻璃的结构维护体

The Yokohama Waterfront Project in the Minato Mirai District contemplates the extension of the city center to establish Yokohama as Japan's second major metropolis in the international marketplace. The complex provides offices, hotels, residential development, the Yokohama Design Port, retail, entertainment, restaurants, a symphony hall, and parking.

Architecturally, the scheme is a composition of bays and towers connected by a liner above a base. The towers, containing the building cores, flank the streets and denote major entries into the retail precincts. Between these towers are the offices, hotels, and residences. The base is trisected by two atria in the east–west direction, providing a city–water connection. It is bisected in the north–south direction by the Queen Mall. The programmatic elements are situated under the wing-like roof, creating a public terrace level between the roof of the individual elements and the wing.

位于未来港区的横滨滨水地带工程希望能扩展城市中心，以使得横滨在国际市场上成为日本的第二大城市。该综合建筑群包括办公楼、旅馆、居住区开发、新设计的横滨滨水区、零售商业区、娱乐区、餐馆、交响乐大厅以及停车场。

在建筑设计上，该方案由多跨开口和塔楼所组成，由一条在裙房之上的线条将它们连接起来。包含着建筑核心筒的塔楼沿街道展开，并暗示着进入零售商业区的主要入口通道。在这些塔楼之间是办公室、旅馆及住宅。裙房被两个中庭沿着西东方向分为三个部分，从而形成了一种城市与水体之间的联系。该方案在南北方向上被女王商场分成两部分。在飞翼状的屋面之下设置了一些建筑中的功能活动要素，在独立要素的屋面和飞翼屋面之间形成了一个公共平台。

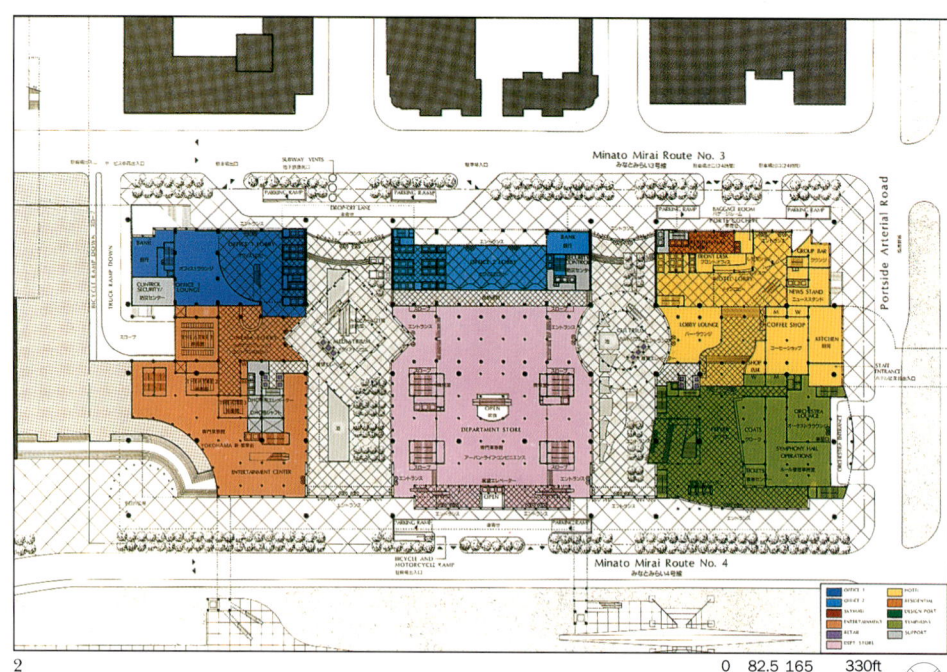

1	Concept sketch	1	构思草图
2	Ground-floor plan	2	一层平面图
3	South-east elevation	3	东南向立面
4	View from the east	4	东向透视

Yokohama Waterfront MM21 Project 87

Tokyo Teleport Town

Design 1990
Tokyo, Japan
Mitsui Real Estate/Kajima Corporation
Associate Architect: Kajima Corporation
900,000 square feet
Structural steel frame with exposed bracing
Glass curtain wall and granite paving

东京特勒波特中心

设计　　1990
日本，东京
三井房地产/鹿岛株式会社
合作者：鹿岛株式会社
900,000平方英尺
钢框架结构及暴露式斜撑
玻璃幕墙及花岗石铺地

The Tokyo Teleport Town extends the Tokyo metropolis onto land reclaimed from Tokyo Bay. The site borders a central promenade that is terminated by the Telecom Center Building.

The scheme comprises two nearly identical towers. By turning one tower 45° towards the other, the composition assumes a certain strength which relates to the Telecom Center yet holds its own. The siting of the two buildings also resolves the symmetrical and asymmetrical qualities of the site. The buildings are simple in form and express their refinement through meticulous detailing. The orthogonal tower is clad in blue glass while the rotated tower is clad in green glass, avoiding the obvious solution of two identical towers.

The towers are positioned on a raised podium which unifies the space between them, establishes a common entry to the offices and amenities, and provides a special space for outdoor activities. The space between the buildings is further defined by a glass roof supported by steel trusses at the thirteenth floor.

东京特勒波特中心将东京市区扩展到了对东京湾进行围海造田所形成的土地上。该建筑基地与一条中央公共散步道接壤，该道路一直延伸到电信中心大楼处结束。

该方案由两个几乎完全相同的塔楼所组成。通过将其中一幢大楼向另一幢旋转45°而构成，该构图在一定程度上体现出某种与电信中心大楼形成联系的力量，同时也保留着自身的特色。两幢塔楼的位置也解决了该基地所具有的对称与不对称的特性。建筑物的造型简单，但它们通过精心的细节处理来展现出其精致的特征。正交布置的塔楼在外墙上用蓝色玻璃，而旋转的塔楼则采用绿色玻璃，从而避免了采用过于明显的方式来区分两座相同的塔楼。

两座塔楼都放置在一个抬高的裙房上，将在它们之间的空间统一起来，并形成了通往办公楼和休闲设施的公共入口，同时也为户外活动提供了一个专门的场所。此外，该方案运用了一个玻璃屋顶来进一步限定在两座塔楼之间的空间，该屋顶设置在第十三层，由钢桁架来支撑。

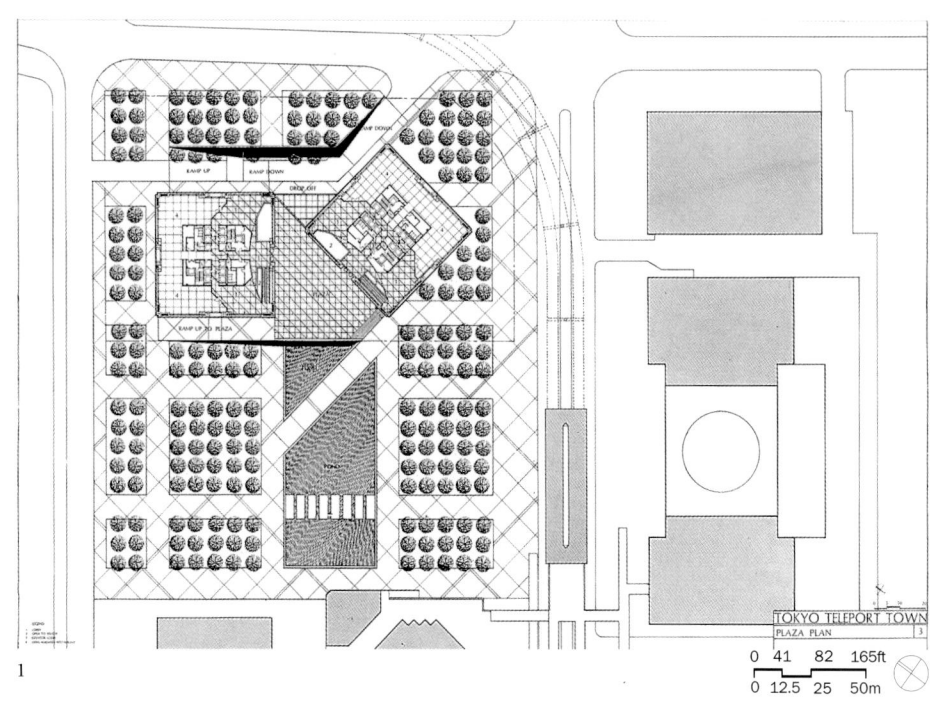

1

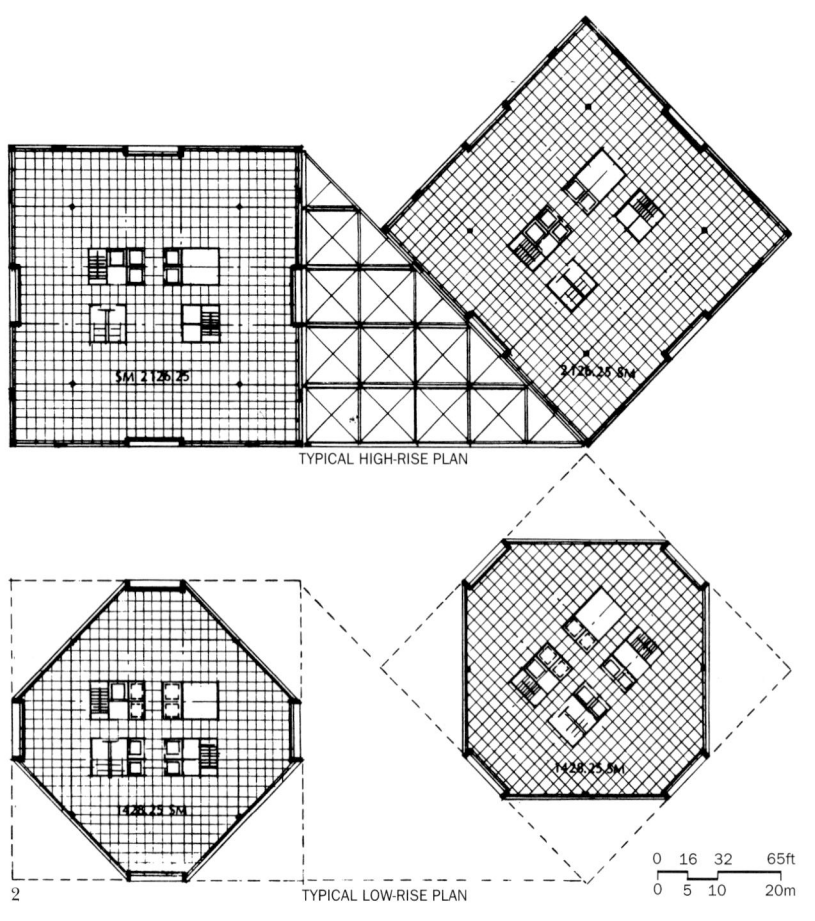

2　　TYPICAL HIGH-RISE PLAN
　　　TYPICAL LOW-RISE PLAN

1 Site plan	1 总平面图
2 Typical floor plans	2 标准层平面图
3 Interplay of rotated buildings	3 旋转的建筑物之间的相互影响
4 Lobby	4 入口大厅
5 Buildings and plaza elements	5 建筑物与广场

3

4

5

425 Lexington Avenue

Design/Completion 1985/1989
New York, New York
Olympia & York Equity Corporation
565,000 square feet
Steel frame, metal deck
Aluminum curtain wall with green and buff granite, insulated vision glass and fritted spandrel glass

莱克星顿大街 425 号

设计/竣工　1985/1989
纽约州，纽约市
奥林匹亚和约克股票公司
565,000 平方英尺
钢框架结构，金属楼面
绿色和浅黄色花岗石及铝材幕墙，不透明玻璃及热熔层间玻璃

The project site occupies a full-block frontage between 43rd and 44th Streets along the eastern side of Lexington Avenue in midtown Manhattan. The features of the landmark Chrysler Building, situated directly across 43rd Street to the south, had a decisive influence on the design concept.

The building typology is an innovative variation on the "architectural column". The plan of the tower is a rectangle with strongly chamfered corners, rising from a stepped and similarly chamfered base. The tower top, like the capital of a column, hovers over the shaft and base and, in a contrasting gesture, flares outward at the same height at which the Chrysler tower steps inward.

The main entrance, on Lexington Avenue, is bounded by retail shopfronts. This entrance is the dominant recess in the streetwall, and leads directly to a "grand salon" type lobby with a coffered, luminous ceiling.

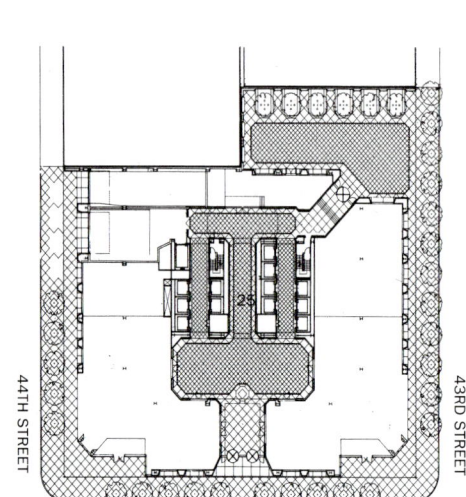

该项目的基地位于第 43 号及第 44 号大街之间的一块整个街区大小的空地上，其正面沿着曼哈顿中心区里的莱克星顿大街的东侧展开。坐落在第 43 号大街南向对面的地标性建筑克莱斯勒大厦的形象对该建筑的设计构思有着决定性的影响。

该建筑在形体上是对"建筑式柱子"进行的一种创新变形。塔楼的平面为一矩形，在角上留出大斜切面，将塔楼放置在一个有退台处理和相似的斜切面的裙房上。塔楼的顶部像是一个柱头，高高地漂浮于塔身和裙房的上方，并用对比的手法使塔楼顶部在克莱斯勒大厦向内退台的相同的高度上向外张开。

位于莱克星顿大街方向上的主入口周边为零售商业店面，该入口是沿街墙面上的主要退进口，人们可由它直接通往一个有藻井式发光顶棚的"大沙龙"型的大厅。

1　底层平面图
2　连接克莱斯勒大街的莱克星顿大街入口处
3　入口大厅
4　幕墙顶部的细部处理

1 Ground-floor plan
2 Lexington Avenue entrance with Chrysler Building
3 Lobby
4 Curtain wall top detail

750 Lexington Avenue

Design/Completion 1984/1989
New York, New York
Cohen Brothers
365,000 square feet
Steel structure
Painted aluminum mullions with tinted and reflective glass; blue pearl granite spandrels and base

莱克星顿大街 750 号

设计/竣工　　1984/1989
纽约州，纽约市
科恩兄弟公司
365,000 平方英尺
钢结构
涂色铝材竖框及有色反光玻璃，
蓝色珍珠花岗石层间贴面及裙房贴面

750 Lexington Avenue is a 30-story office tower occupying a key site within midtown Manhattan. It is directly on top of a major subway station, and major improvements to the subway entrances are included as part of the project.

The design involves the interlocking of two geometric volumes: an inner cylinder is wrapped by an outer volume which changes its configuration from a rectangle to a polygon based on 45° geometry. The cylinder projects from this form at the top, and also on Lexington Avenue to form a giant bay window. The stepped top, with its projecting light feature, gives the building an interesting profile and promotes recognition.

On the ground floor along Lexington Avenue the retail frontage extends beyond the outline of the tower to reinforce streetwall continuity and provide retail exposure. Along the western property line, similar extensions serve as office building entry points.

　　莱克星顿大街 750 号是位于曼哈顿中心闹市区一块重要基地上的一栋 30 层高的办公大楼。该建筑物直接坐落在一个重要的地铁站的上方，因此，对地铁入口进行大的改进也就成为了这一项目的一个组成部分。

　　该项目在建筑设计上为两个相互连接起来的几何体量：内部的圆柱体被一个外部体量所包围，使其从一个矩形变为基于 45°斜角的多边形构图。在整个造型的顶部，圆柱体从这一构图中凸显出来，并且在莱克星顿大街方向上形成了一个巨大的凸窗。突出的轻灵的顶部进行了退台式的处理，使得大楼轮廓生动而有趣，从而强化了其可识别性。

　　在沿莱克星顿大街的一层上，临街的商业零售门面延伸到塔楼外轮廓以内，强化了沿街立面的连续性，也将零售部分展示出来。沿建筑红线的西侧，采用了相似的延伸方式来作为办公大楼的入口处理。

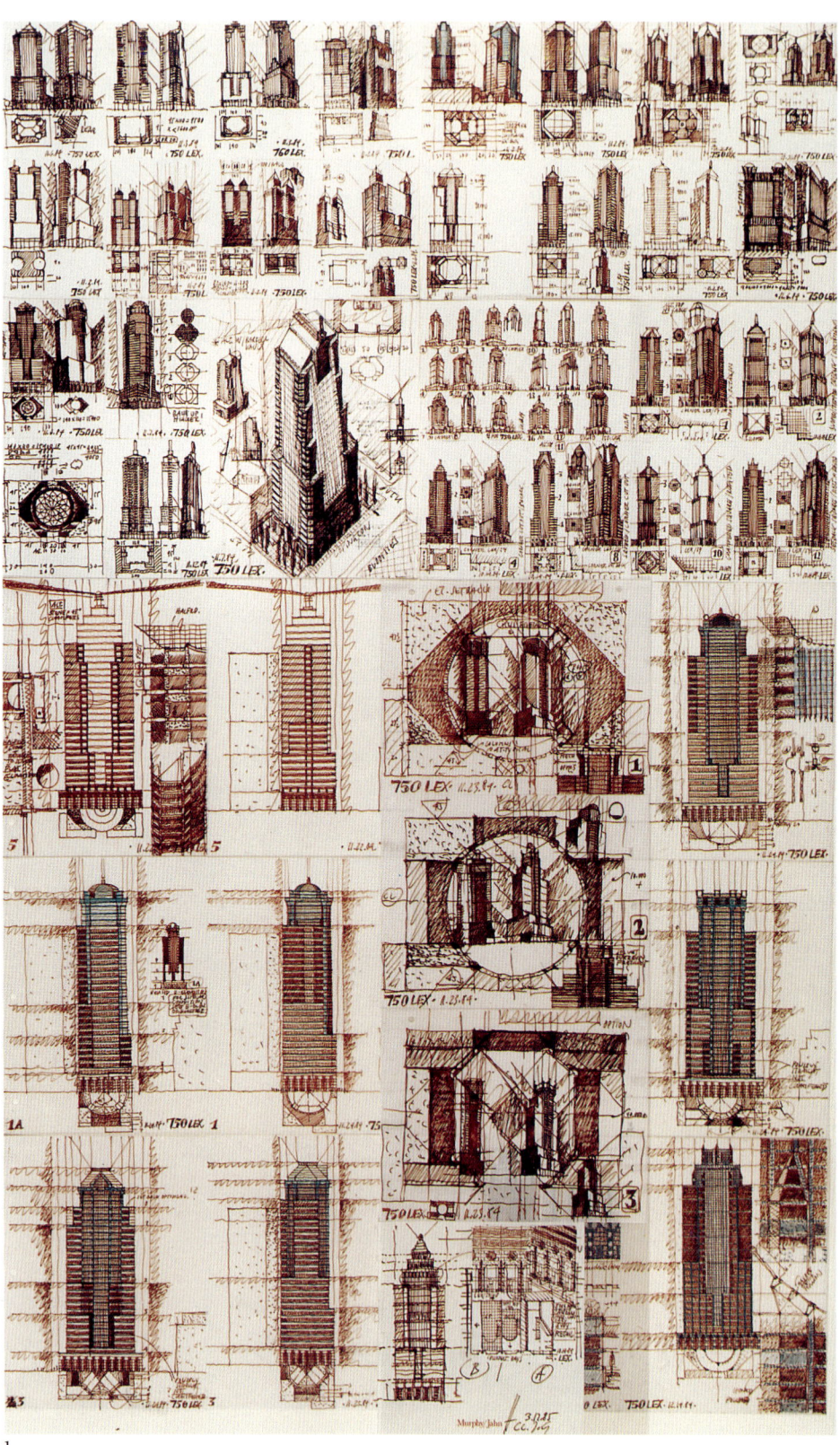

1

1 Concept sketches
2 Stepped top
3 City context
4 Interlocking skin elements
5 Streetscape showing retail

1 构思草图
2 退台式的顶部处理
3 城市周边环境
4 互相影响的外墙元素
5 街道景观上的商业门面

2

3

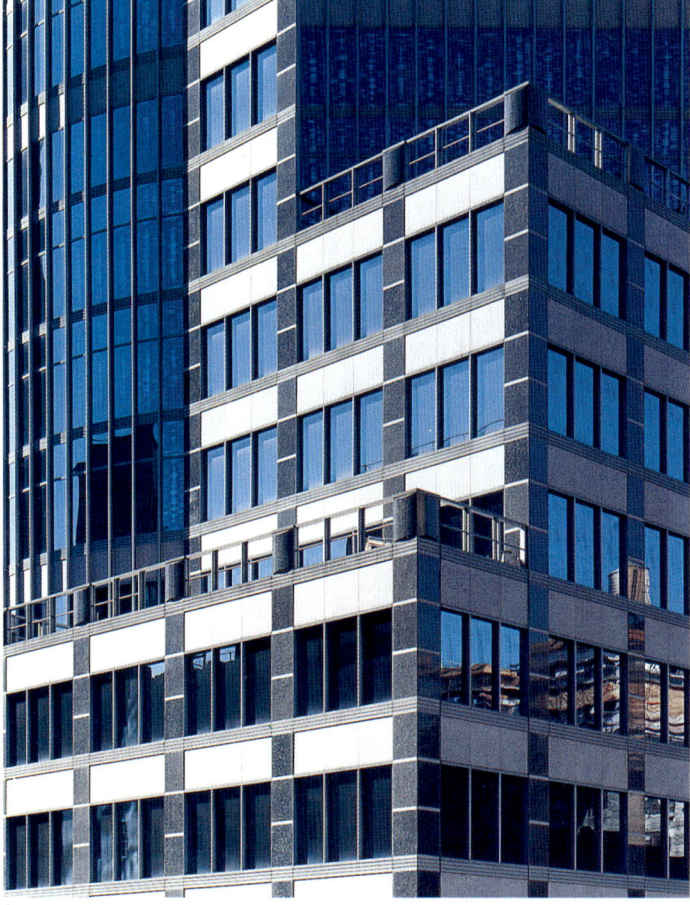
4

5

Cityspire

Design/Completion 1984/1989
New York, New York
West 56th Street Associates
842,000 square feet
High strength poured-in-place concrete
Tinted vision glass, vertical and horizontal granite panels, painted aluminum mullions, painted steel plate ribs and aluminum fins at the dome

城市之颠

设计/竣工　　1984/1989
纽约州，纽约市
西56号大街联合会
842,000平方英尺
高强度现浇混凝土
彩色观景玻璃，垂直及水平花岗石板贴面
涂层铝材竖框，涂层钢板肋架及穹顶部分的铝制翼片板

Situated at the northern edge of New York City's historic Broadway Theater District, the Cityspire rises as one of Manhattan's tallest and most prominent residential/mixed-use towers.

The site is adjacent to the recently landmark-designated City Center Theater of Music and Drama. A transfer of unused development rights above the City Center Theater enabled this mid-block project to reach a height of approximately 70 stories, with 353 luxury condominium units surmounting 23 stories of speculative office space and retail.

Careful attention to scale and materials at the new tower's base attempts to relate the 56th Street facade to its existing neighbor. Rising out of this base in a series of three setbacks is a stone-clad octagonal shaft with lateral wings of glass projecting to the east and west. These wings ultimately terminate at the sixty-first floor, releasing an 80-foot diameter omni-directional tower with copper dome into the skyline.

"城市之颠"建筑位于纽约市具有历史意义的百老汇剧院区的北部边缘地带，它高高耸立，是曼哈顿岛上最高的、最显著的居住和多功能塔楼建筑之一。

该用地紧挨着近来被看作是地标性建筑的城市音乐与戏剧中心剧院。开发者将城市中心剧院未使用的上空开发权转移过来，使得这个位于地块中部的项目能够达到约70层楼的高度，拥有353个豪华公寓单元，下面有23层的办公及商业零售空间。

这个新塔楼的裙房部分在尺度和材料的使用上非常小心，希望将沿着56号大街的立面与其他相邻的建筑形成联系。在该裙房之上，耸立着一个八边形的柱形体量，有3层退台，采用石材贴面，其东西两侧各伸出一个横向的玻璃带状的侧翼。这些侧翼最终在第六十一层处结束，在其上面是一个全方向性的塔楼，直径为80英尺，顶部的铜制穹顶形成其天际轮廓线。

1　无方向性的塔楼，顶部是铜制穹顶

1　Omni-directional tower with copper dome

Project J (1, 2, 3)

Design 1989
Chicago, Illinois
Howard Ecker & Co.
PJ-1/PJ-2: 600,000 square feet
PJ-3: 400,000 square feet
Slip form concrete core
Poured-in-place column and slab, stone and aluminum skin

These projects represent similar concepts developed for a number of sites and sizes.

PJ-1 is an office building that develops from the overlapping and integration of multiple architectural elements, and the influences of the program, the neighborhood, and current technology.

Typologically the building is a corner building that responds to the urban situation of the city to the south and the lake to the east. The corner is clearly articulated as a round bay with a recessed entry at street level and full circular floors at the building top.

PJ-2 is an office building with retail and parking, located on the near north side of Chicago. The goal was to create a state-of-the-art building of superior quality to attract first-rate tenants. PJ-3 was developed for an alternative site north of the river, and is essentially a downscaling of PJ-2 in anticipation of reduced market conditions, while still being a highly visible building.

1

2

3

1 Project J-1
2 Project J-2
3 Project J-3: concept sketches

Wilshire/Westwood

Design/Completion 1984/1988
Los Angeles, California
Platt Corporation
230,000 square feet
Ductile steel moment frame
Color-coated aluminum window wall system with reflective gray glass;
Kasota limestone, granite on strong back truss system

The building is located along a section of Wilshire Boulevard lined by typical high-rise development. The form of these buildings is reduced to a slab, square or octagon, with the skin representing the horizontal layering of office floors, the verticality of the columns, or the structural bays. The design of this project attempts to collage these "fragments" of the urban scene into a new form. A rectangular slab with sloping and stepped ends is combined with two octagonal towers that read as giant columns. A combination of glass and stone finishes accentuate the setbacks and create a capital-like top.

At the street level, the building abuts the property lines with commercial uses, creating an arcaded appearance. A large octagonal public reception space at the corner serves as a monumental entry. The remainder of the site has been landscaped as a public park to make the transition to the residential scale to the south.

威尔希尔/威斯特伍德

设计/竣工　1984/1988
加利福尼亚州，洛杉矶市
普拉特公司
230,000平方英尺
韧性钢弯矩框架结构
彩色涂层铝材窗式墙体系统及反射式灰色玻璃
坚固的背面桁架系统上覆盖卡索塔石灰石、花岗石

大楼位于威尔希尔林荫大道的一段上，沿着该大道并列着一些典型的高层建筑。这些建筑物的造型可以简化为一个板式建筑、一个正方形或八边形建筑，外墙表现出办公楼的水平条带的特征、柱子的垂直排列、或是结构开间的表现。该项目在设计上尝试将这些都市风貌的"片段"进行拼贴组合，形成一种新的形式。方案将一个有着倾斜的并呈阶梯形的末端的矩形大板体量与看上去宛如巨大的柱子的两个八边形塔楼组合在一起。玻璃与石材装修的组合加强了建筑体量的凸凹关系，并形成了一种柱头式的顶部效果。

在平街层上，该建筑将商业功能部分毗邻建筑红线布置，形成一种柱廊式的外观。一个位于角落上的巨大的八边形公共接待空间被用作纪念性的入口。该工程基地的剩余部分做了绿化处理，形成一个供公众使用的公园，以使该建筑与在南面的居住建筑的尺度之间形成过渡。

1　构思草图
2　总平面图
3　威尔希尔街与米德维尔街转角处的立面

1　Concept sketches
2　Site plan
3　Wilshire/Midvale elevation

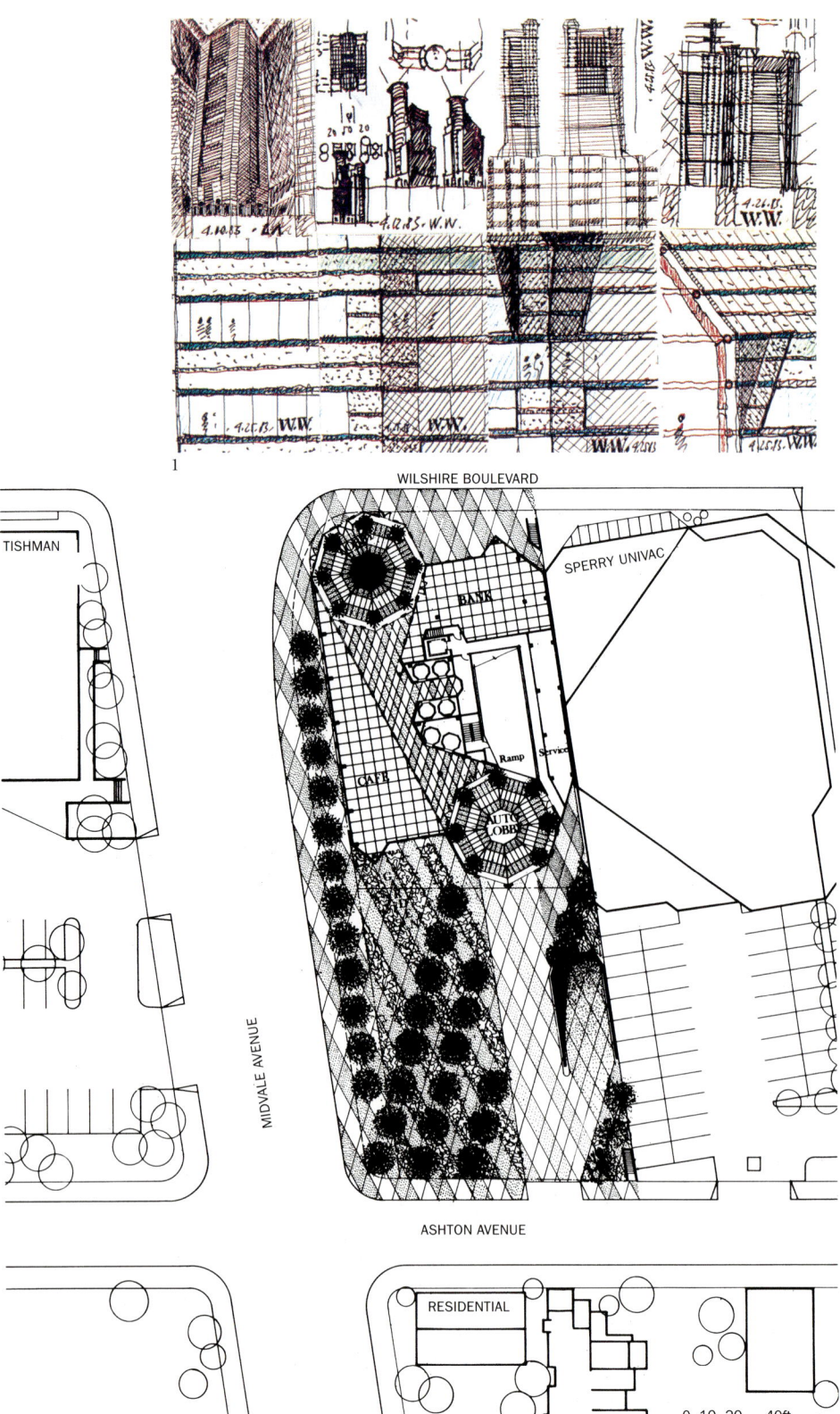

4	Car entrance		4	汽车入口
5	Typical floor plans		5	标准层平面图
6	Midvale elevation		6	米德维尔街立面
7	Lobby		7	入口门厅

4

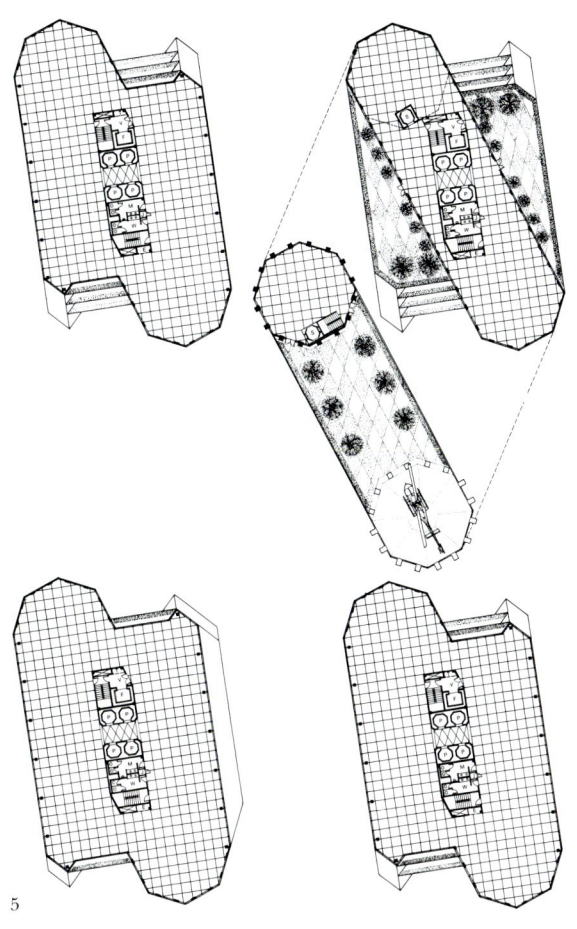

5

6

7

Wilshire/Westwood 99

1111 Brickell Avenue

Design 1988
Miami, Florida
Brickell Associates
1,740,000 square feet
Steel and concrete structure
Glass and aluminum curtain wall

布里克尔大街 1111 号

设计　　1988
佛罗里达州，迈阿密市
布里克尔合伙人公司
钢与混凝土结构
玻璃与铝材幕墙

1111 Brickell Avenue is a mixed-use complex consisting of two high-rise towers sitting on a base fronting a public plaza. One tower is exclusively for office use. The upper portion of the second tower provides 230 hotel rooms and associated public areas, with office space on the floors below. The base houses the entry lobbies for the office and hotel functions; a two-level retail arcade; parking for 1,150 cars; and building services.

Each tower is sculptured differently to give it an identifiable form. The tower facing towards the south-west corner of Brickell Avenue has a cantilevered, semi-circular form that provides panoramic views. The second tower has two cantilevered rectangular bays that meet in a re-entrant corner facing Brickell Avenue. The two towers and the base are tied together by a two-story, 30-foot grid of white marble or granite, with black granite at the street level and between the towers.

布里克尔大街 1111 号是由两栋高层塔楼组成的多功能综合建筑，其底部裙房面向一个公共的广场。其中的一个塔楼只用于办公。另一个塔楼的上面部分设置了 230 个旅馆客房及相关配套的公共区域，下面部分是办公空间。裙房部分设置了办公和旅馆部分的入口大厅、一个有两层楼高的商业零售廊道、可停放 1,150 辆车的停车场以及建筑的服务性设施。

两个塔楼采用了不同的处理手法，形成各自明确的形象特征。朝向布里克尔大街西南角的塔楼上设计了一个悬挑式的半圆形体，人们可以在此观赏到城市的全景。另一个塔楼也有两处悬挑出来的矩形体量，两者在面向布里克尔大街的一个凹角处连接起来。两个塔楼及裙房由两层高 30 英尺见方的白色大理石或花岗石网格来进行统一，在平街层和两个塔楼之间的地方饰以黑色花岗石。

1　South elevation
2　View from the south-west
3　West elevation

1　南向立面
2　从西南方向看建筑
3　西向立面

The North Loop Block 37

Design 1988
Chicago, Illinois
FJV Venture
2,300,000 square feet
Concrete structure
Granite and glass curtain wall

北环城路第 37 号街区

设计　1988
伊利诺伊州，芝加哥市
FJV 风险投资公司
2,300,000 平方英尺
混凝土结构
花岗石与玻璃幕墙

Ambitious in scale as well as intent, this project confirms the latent vitality of Chicago's North Loop, providing a vision of the future for this part of the city. The project is conceived as a full-block development consisting of 300,000 square feet of retail and two office towers, organized along a pedestrian arcade.

The backbone of the project is the 320-foot four-story retail arcade with central rotunda, which connects the office population to the west with existing retail activities to the east. It offers a protected urban space accessible only to pedestrians, and provides connections between essential points of origin and destination.

North and south of the arcade, the two office towers rise above the retail activities providing 1.9 million square feet of office space. Each building has a separate corner entry along Dearborn Street, while the second-story lobbies are linked directly to the arcade.

　　该项目在规模上和设计意图上都显得雄心勃勃，从而显示了芝加哥北环城路所具有的潜在活力，并为芝加哥市的这一地区的未来展现出一个美好的前景。该项目在构思上考虑将整个街区进行开发，其中包括500,000 平方英尺的商业零售区和两栋高层办公塔楼，将它们沿一个步行廊道来进行组织。

　　该项目的中枢部分是高 320 英尺的带中心圆形大厅的 4 层商业零售廊道，这一部分将西侧的办公人群与东侧现有的商业零售活动连接在一起。它提供了一个仅供行人进入的受到庇护的城市空间，并且在主要的出发点与目的地之间形成某种连接。在廊道的南北两侧，两栋塔楼从商业零售区上延伸出去，形成了 190 万平方英尺的办公空间。在迪尔伯恩大街上，每栋建筑都有着各自独立的在转角处的入口，而在第二层上的入口大厅直接与商业廊道相连接。

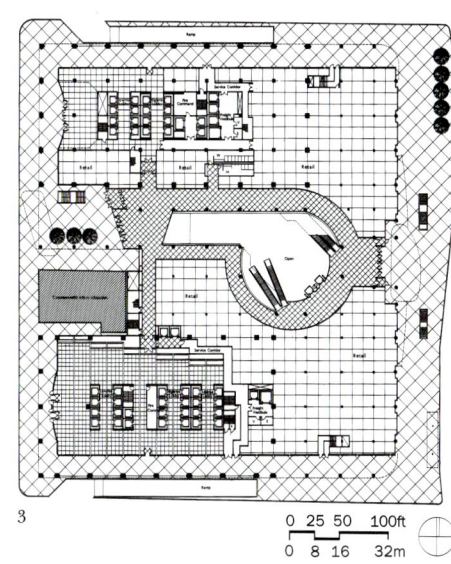

1　位于迪尔伯恩大街上的市民中心广场
2　州立大街立面
3　底层平面图

1　Dearborn Street with Civic Center Plaza
2　State Street
3　Ground-floor plan

Oakbrook Terrace Tower

Design/Completion 1985/1987
Oakbrook Terrace, Illinois
Miglin Beitler Inc.
714,000 square feet
Reinforced concrete core and steel frame
Aluminum and glass curtain wall; terrazzo floors with marble cladding in lobby

奥克布鲁克台地高层塔楼

设计/竣工　1985/1987
伊利诺伊州，奥克布鲁克台地
米格林·拜特勒公司
714,000 平方英尺
钢筋混凝土核心筒及钢结构
铝材和玻璃幕墙，水磨石地面，大理石饰面

The Oakbrook Terrace Tower provides 663,000 square feet of office space in a 29-story tower. The building is sited on 16 acres in the suburb of Oakbrook Terrace approximately 25 miles west of Chicago.

The tower is conceived as an octagon in plan, with emphasis on the orthogonal planes. The four diagonal faces employ a horizontal expression that, when played against the vertical, provides a high level of visual interest from the surrounding road network. This play of horizontal and vertical generates the tower's top which is the building's signature.

The tower provides typical floors of 23,800 square feet. The twenty-ninth floor contains an additional mezzanine, providing a unique opportunity for a tenant who requires a distinct identity. The lobby atria reinforce the ceremonial aspects of the design. A restaurant is located at the ground floor, while the lower level contains a health club as well as parking for 200 cars.

奥克布鲁克台地的高层塔楼有29层，可提供663,000平方英尺的办公空间。该建筑位于奥克布鲁台地郊区的一块16英亩的基地上，在芝加哥西面约25英里的地方。

塔楼的平面被设计成八边形，重点强调其正交的水平面。四个对角线方向的外墙面采用水平表现的手法，与垂直表现形成对比，使人们从其周围的道路上可以获得高水平的视觉兴趣。这种水平与垂直对比的建筑手法构成了塔楼的顶部，并成为该建筑的标志。

塔楼有23,8000平方英尺的标准层建筑面积。第二十九层上还设置了一个附加的夹层，为追求个性的租赁户提供了一个独特的机会。大厅的中庭增强了该设计所具有的纪念性风格特征。在底层上设置了一个餐厅，在其地下层中布置了一个健身俱乐部和可停放200辆汽车的停车场。

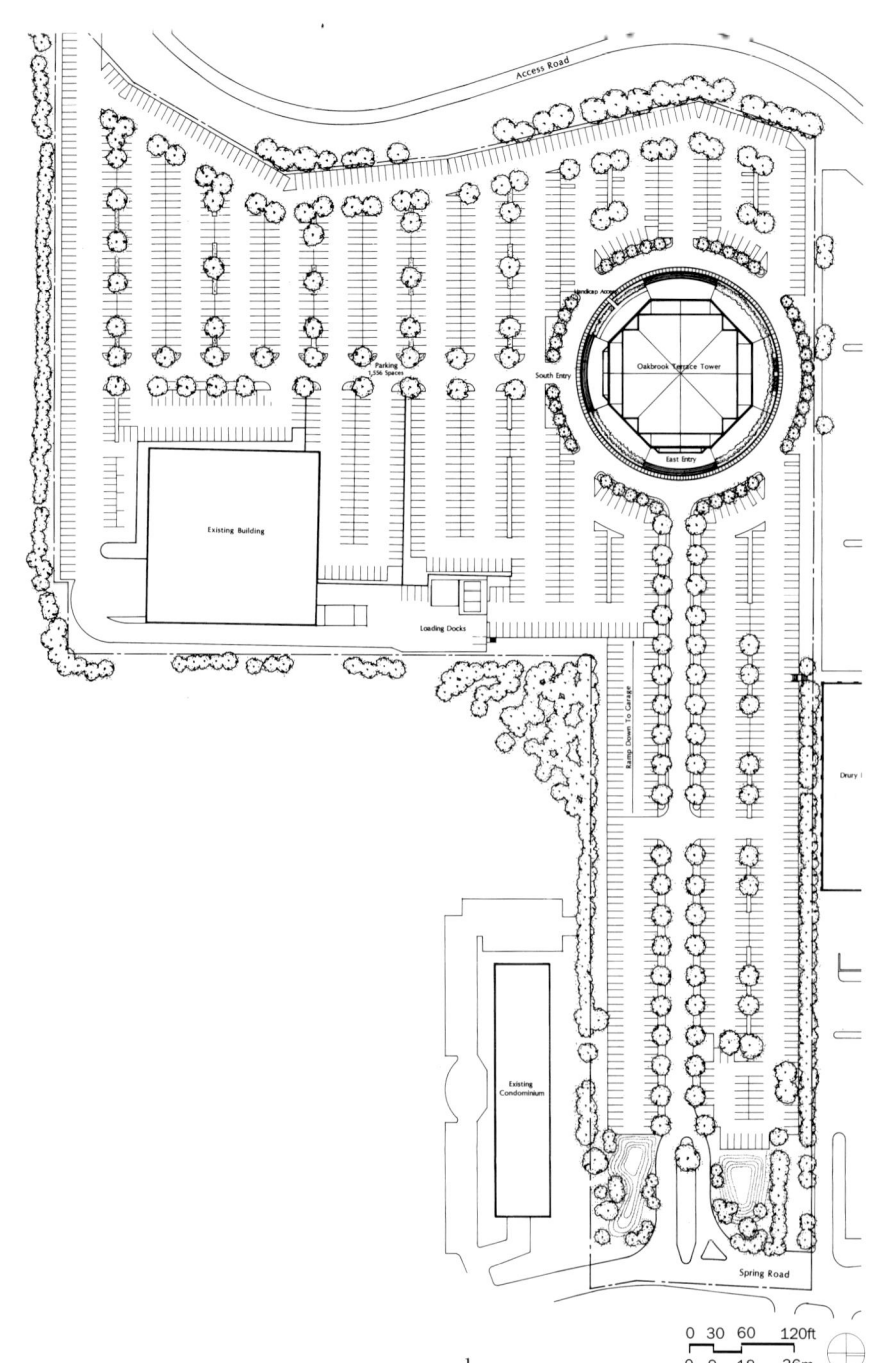

1　Site plan
2　Vertical entrance facade with horizontal chamfered corners
3　Entrance
4　Lobby
5　Elevator lobby

1　总平面图
2　垂直处理的入口立面及水平式的斜切转角
3　入口
4　入口门厅
5　电梯间

2

3

4

5

Oakbrook Terrace Tower 103

One Liberty Place

Design/Completion 1984/1987
Philadelphia, Pennsylvania
Rouse Associates
1,100,000 square feet
Steel frame and composite deck
Fluoropolymer-coated aluminum curtain wall with reflective glass, polished and honed granite

自由城一号

设计／竣工　1984/1987
宾夕法尼亚州，费城
劳斯合伙人公司
1,100,000平方英尺
钢框架结构及混合式楼板
含氟聚合物涂层铝材幕墙及反射玻璃、抛光和细磨花岗石

The Liberty Place project occupies a prime block in central Philadelphia, assembled through the closing of two streets. One Liberty Place is the first phase of the project, sited on the corner of 17th and Market Streets, and contains 61 stories of office space. Special high-rise zoning allows the project to exceed the 491-foot height limit of William Penn's statue on top of the Philadelphia City Hall.

The tower itself is square with re-entrant corners. The shift in plane between the building's shaft and the corners emphasizes the location of eight "super columns" which are tied with outriggers to the building's braced central core at intermittent floor levels. This "outrigger/super column" structure permits greater flexibility and openness for office space and ground-floor retail than the conventional "tube" system.

From the base, which is clad in stone, the amount of glass increases until the top is totally sheathed in glass.

自由城项目位于费城市中心区的一个重要的街区上，该街区由两条大街相交而形成。自由城一号是该项目的第一期，位于第17大街及市场大街的转角处，设计了61层的办公空间。专门进行的高层建筑分区设计允许该项目在高度上超过设置在费城市政厅顶上的威廉·潘雕像所设定的491英尺的限制高度。

该塔楼本身呈正方形，设有凹入式的转角处理。在塔楼主体和转角处之间所形成的平面上的变化强调了八根"超级柱子"所在的位置，并通过悬臂梁以隔层设置的方式将这些柱子与建筑的带斜撑的中央筒体连接起来。这种"悬臂梁与超级柱子"所组成的结构与传统的筒体结构相比，可使办公空间和底层上的商业零售区获得更大的灵活性和开敞性特征。

从采用石材贴面的裙房部分开始，玻璃的使用量逐渐增多，建筑物的顶部则完全被包裹在玻璃里了。

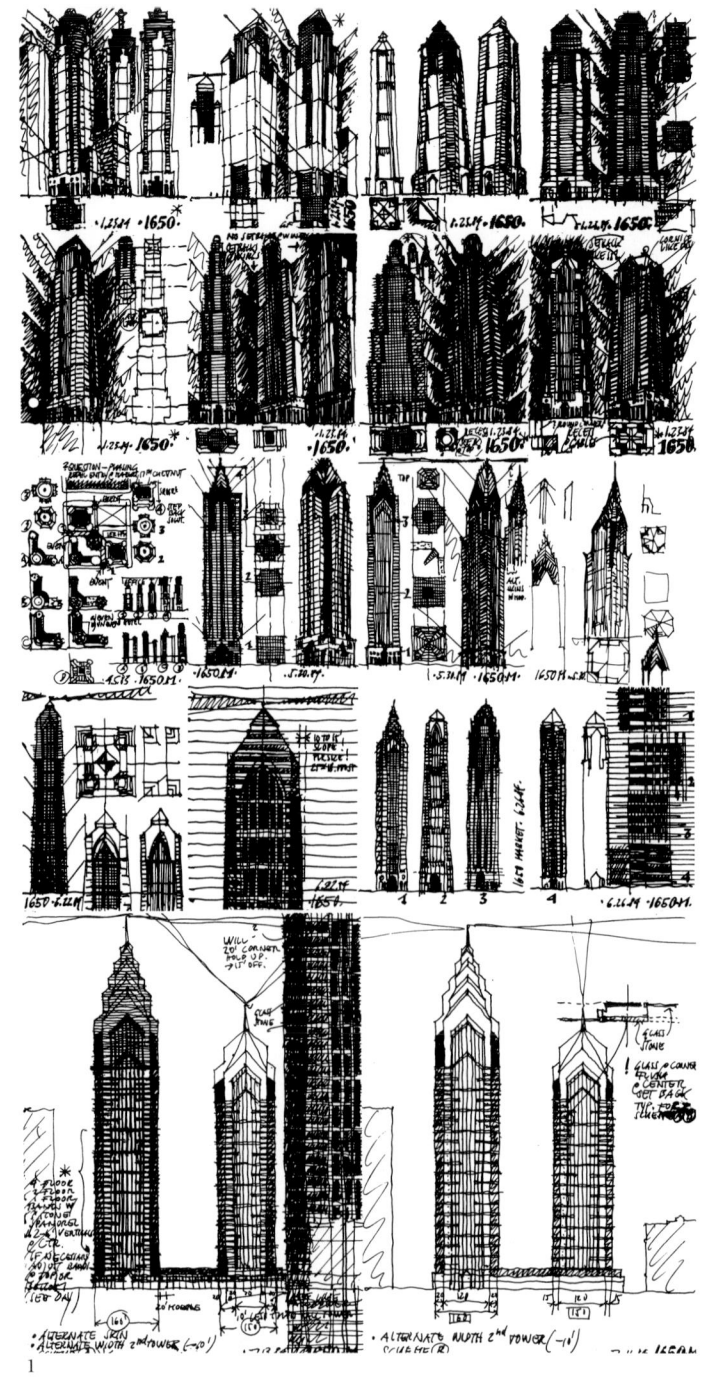

1

1	构思草图	1	Concept sketches
2	城市景观	2	Cityscape
3	全景：自由城一号	3	Overall view: One Liberty Place
4	自由城Ⅰ期和Ⅱ期	4	Liberty Place phases I and II

2

3

4

One Liberty Place 105

5

6

7

5	Entrance	5	入口
6	Side entrance	6	侧入口
7	Streetscape with storefronts	7	街道景观及临街店面
8	Lobby	8	入口门厅

Northwestern Atrium Center

Design/Completion 1984/1987
Chicago, Illinois
Tishman Midwest Management Corp.
1,600,000 square feet
Braced perimeter steel frame; atrium framing of lightweight steel trusses
Exterior tube aluminum curtain wall with infill of reflective silver, reflective blue, and black-tinted glass

西北中庭中心

设计/竣工　　1984/1987
伊利诺伊州，芝加哥市
提希曼中西部管理公司
1,600,000平方英尺
斜撑式外围钢结构，轻质钢桁架的中庭构架
外筒为铝材幕墙，镶嵌反光银色玻璃、反光蓝色玻璃、以及黑色玻璃

The Northwestern Atrium Center Project is a combined commuter terminal and office building complex replacing the existing Chicago and Northwestern Train Station. The building forms a major gateway between the station and Madison Street, removing the "Chinese Wall" effect of the existing station on the growth of the central business district.

A continuous arcade along Madison Street leads to a sequence of multi-story atria which enhance commuter traffic flow to the train platform. The atrium spaces bring light to the inside without the penalty of heat loss or gain.

On street and track levels, 85,000 square feet of terminal facilities serve the requirements of a modern commuter station. The office tower is served from a skylobby one floor above track level. The form of the tower reflects the symmetry of the scheme and the varying floor sizes. Its smooth, streamlined curves create a shape which appropriately symbolizes the image of trains.

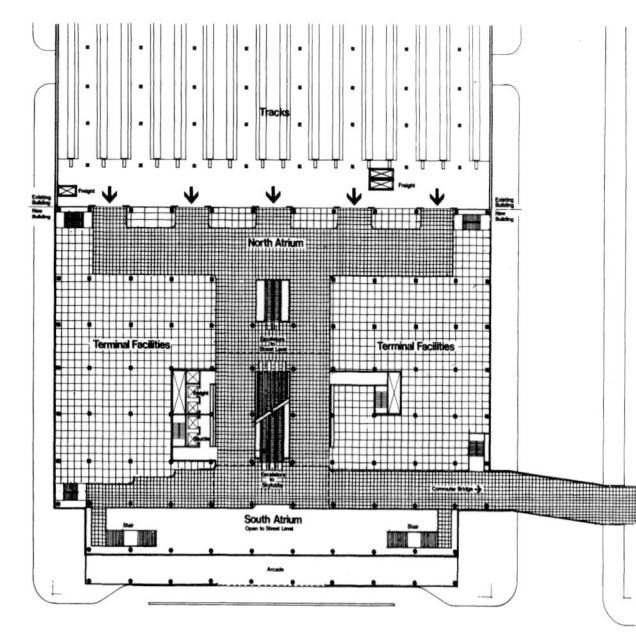

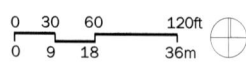

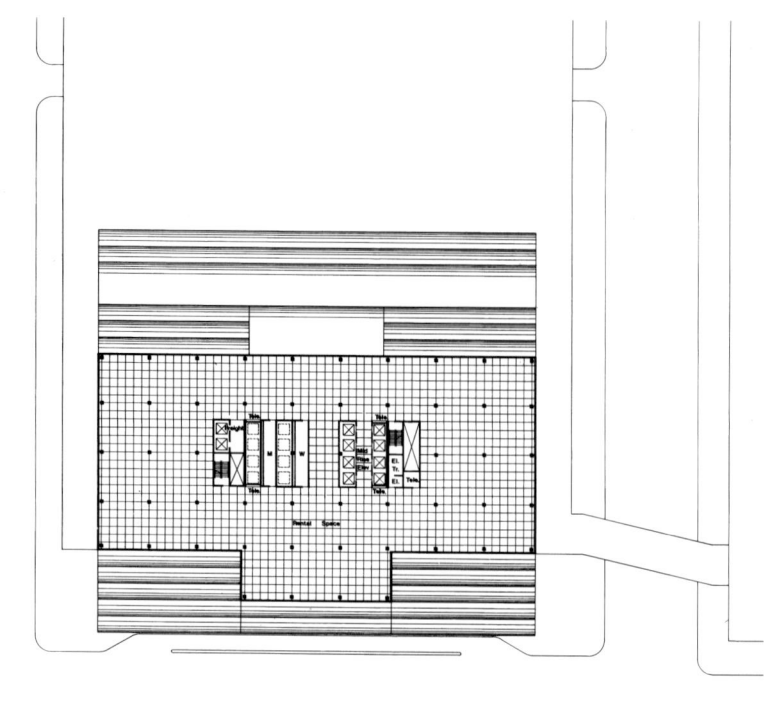

　　西北中庭中心项目是一个将通勤终点站与办公大楼组合在一起的综合性建筑物，用它来替代现有的芝加哥与西北火车站。该建筑在车站与麦迪逊大街之间形成一个重要的通道，以之来消除原有车站上所具有的"中国墙"对这个中央商务区的发展所带来的影响。

　　沿着麦迪逊大街的一条连续的廊道可以进入一系列的多层式中庭，这使得通勤乘客人流可以更快地到达火车站台上。中庭空间将光线引入到建筑内部，同时也能保证没有热量的流失或增加。

　　在街道和轨道平面上，85,000平方英尺的终点站设施可以满足一个现代化的车站的需求。轨道平面的上面一层有一个空中大堂，为上方的办公塔楼提供服务。塔楼的外形反映出设计上的对称性及不同楼层在面积上的变化。它那光滑新颖的曲线所形成的造型恰如其分地象征着火车的形象。

1 Ground-floor plan
2 Typical floor plan
3 South facade
4 Roll-back entrance arcade
5 South atrium

1 底层平面图
2 标准层平面图
3 南向立面
4 层叠缩进的入口廊道
5 南向中庭

3

4

5

Northwestern Atrium Center 109

6 North atrium passenger train arrival/waiting area
7 South entrance atrium/retail
8 Escalator to third-floor skylobby
9 Entrance gateway
10 North elevation with passenger train sheds

6 北向中庭旅客列车到达及等候大厅
7 南向入口中庭与商业零售设施
8 通往第三层空中大堂的自动扶梯
9 入口通道
10 北向立面及旅客列车棚

6

7

8

9

362 West Street

Design/Completion 1983/1986
Durban, South Africa
Anglo-American Properties
Associate Architect: Stauch Vorster + Partners
22,000 square feet
Tower: concrete frame with aluminum and glass curtain wall
Arcade: exposed steel frame with glass curtain wall

西大街 362 号

设计/竣工　　1983/1986
南非，德班市
盎格鲁—美利坚房地产公司
合作建筑设计单位：斯道茨·弗斯特及合伙人事务所
22,000 平方英尺
塔楼：混凝土框架结构，铝与玻璃幕墙
廊道：暴露式钢结构，玻璃幕墙

362 West Street is a mixed-use development containing 231,400 square feet of office space in a 24-story tower and 32,300 square feet of retail space in a 2-story base structure. The development provides a continuation of the lively character of the existing streetscape, with shopfronts along the sidewalks as well as retail arcades at mid-block.

The tower is a geometric configuration of two concentric octagons. The outer form spirals about the inner form which culminates in a steep spire housing microwave and communications equipment. A landscaped roof terrace is provided at each floor level, introducing a natural element and providing vistas over the bay and the Indian Ocean.

The physical configuration of this design provides an interesting alternative to the typical office building of identical stacked floors, with varying amounts of floorspace provided at each level.

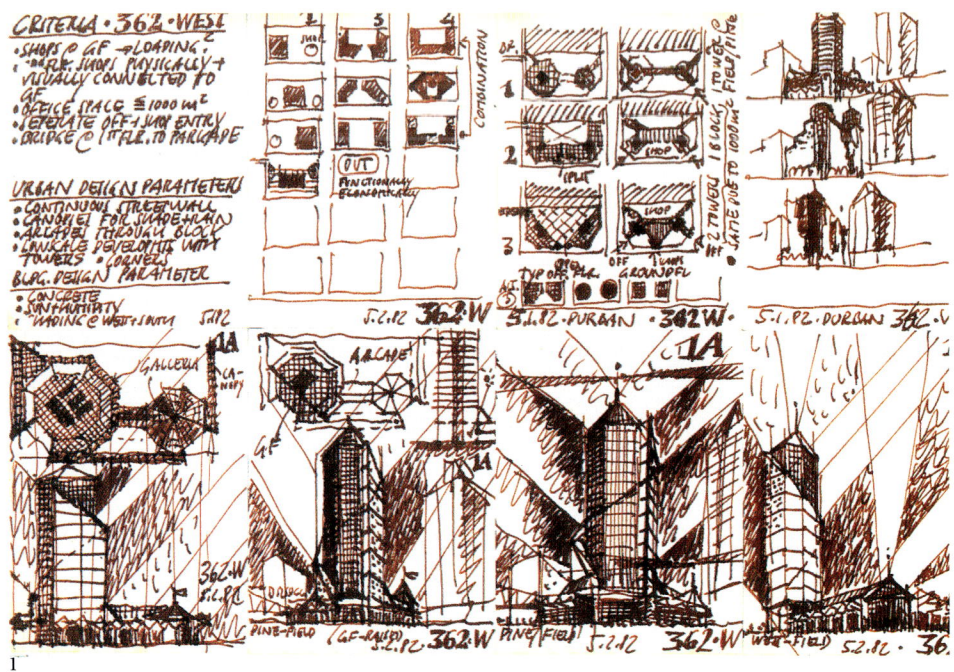

1

西大街 362 号是一个多功能的开发项目，其中包括了一个高 24 层的塔楼里的 231,400 平方英尺的办公空间以及在一个有两层高的裙房里的 32,300 平方英尺商业零售空间。该项目促使已有的街道景观所具有的生动活泼的特征得以延续，沿人行道布置了商业店面，而在街区地块的中段设置了商业零售廊道。

塔楼在几何构图上由两个同心的八边形组成。其外层形体围绕着内层体量盘旋而上，最后是一个高耸的尖顶，其中安装了微波和通信设备。每层楼都有一处带绿化的屋面平台，从而将自然元素引入到建筑之中，并能够观赏到海湾及印度洋远景。

该设计所形成的实体构图为由简单叠加楼层所形成的典型办公建筑提供了一种富有趣味的变化，并且各个楼层的建筑面积在不断地发生变化。

2

1–2　Concept sketches
3　　Typical floor plan
4　　Retail arcade/pedestrian way
5　　Outer and inner octagons with terraces

1–2　构思草图
3　　标准层平面图
4　　商业零售廊道与人行通道
5　　外层和内层的八边形体及平台

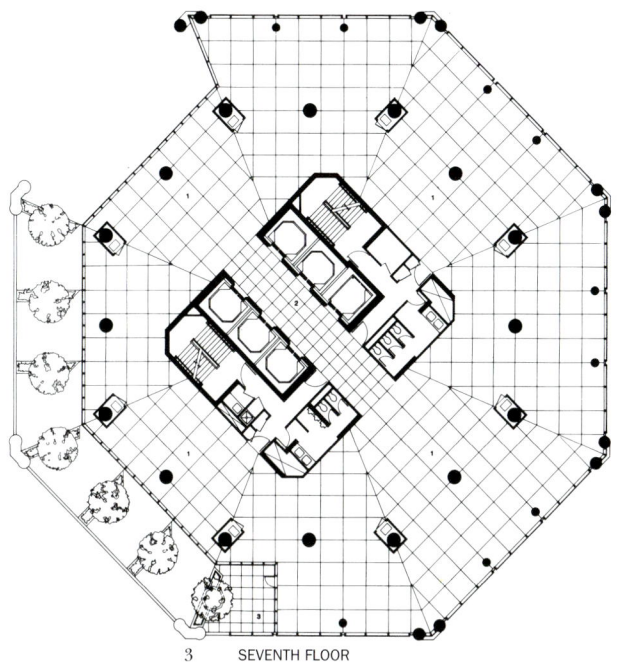
3   SEVENTH FLOOR

4

5

362 West Street   113

Park Avenue Tower

Design/Completion 1982/1986
New York, New York
Park Tower Realty
480,000 square feet
Steel frame with metal deck composite floor construction
Aluminum and glass curtain wall with granite panels glazed into the curtain wall framing system

公园大街塔楼

设计/竣工　　1982/1986
纽约州，纽约市
公园塔楼房地产开发公司
480,000 平方英尺
钢框架结构，金属平台复合楼板
铝材与玻璃幕墙，花岗石板镶入幕墙框架体系

The project occupies a through-block site connecting 55th and 56th Streets between Park and Madison Avenues in mid-town Manhattan. Two distinguished corporate neighbors are the high-rise headquarters buildings for ATT and IBM.

The dominant image projected by the building is that of a monumental granite and glass obelisk. This form resulted from the programmatic requirement for variable floor sizes together with the converging sky exposure curves. The north and south facades of the tower slope inwards eventually converging with the vertical east–west facades to form a perfect 110-foot square at the top. The sloping and chamfered tower walls incorporate the characteristic New York setback in three tiers concealed within the flush streamline taper of the overall form.

The building contains 35 office floors ranging in size from 17,000 gross square feet in the base to an average of 13,500 square feet in the tapering floors of the tower. An urban plaza is located along the entire site frontage of 55th Street.

　　该项目地处曼哈顿岛闹市中心区，基地位于在公园与麦迪逊大街之间的连接着第55号大街与第56号大街的一个完整的街区上。邻近的两栋建筑是AT&T和IBM的高层总部大楼。

　　该建筑所表现出来的主体形象是一个纪念碑式的由花岗石和玻璃所构成的方尖碑。这种造型是根据项目计划要求而得来的，它需要不同的楼面建筑面积和收束式的天际轮廓线。塔楼的北向和南向立面向内倾斜，最终在大楼的顶部与垂直的西向和东向立面汇合，形成一个110英尺见方的正方形平面。倾斜的和切角的塔楼外墙部分采用了充满纽约特色的三段式退台处理，同时将其隐藏于塔楼在整体造型上所采用的平齐的流线型锥体之中。

　　该建筑设置了35层的办公空间，各层的面积有所变化，裙房基座部分的总建筑面积为17,000平方英尺，锥体塔楼部分的楼层面积平均为135,000平方英尺。在沿着整个基地面向第55号大街的地方设置了一个城市广场。

1

1　View from Park Avenue
2　Typical floor plan
3　Entrance
4　Chamfered corner/setback
5　Elevator lobby

1　从公园大街方向看建筑
2　标准层平面图
3　入口
4　斜切边角及退台处理
5　电梯厅

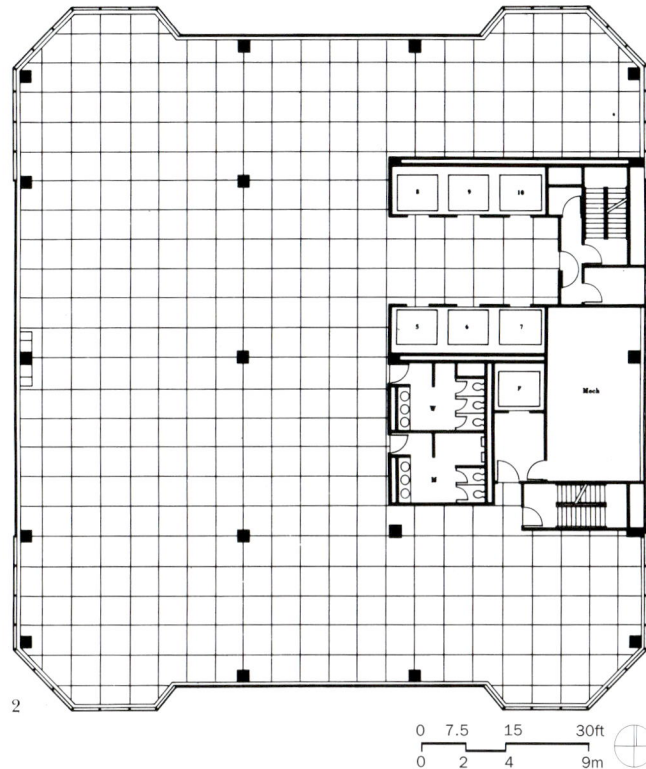

Park Avenue Tower　115

Metro West

Design/Completion 1984/1986
Naperville, Illinois
Westminister Corporation
232,000 square feet
Steel frame structure
Color-coated unitized curtain wall system with reflective gray glass and flamed Luna Pearl light gray granite

Metro West is a 10-story speculative office building located in an office/research park west of metropolitan Chicago. The project consists of two buildings in phased development linked by an enclosed atrium structure. The design establishes a raised landscaped podium for the base of the integrated structures.

The cubic forms of the office buildings are related at a 45° angle on the axis of the platform area. The curvilinear atrium structure is suspended from the opposing faces of the buildings and becomes an independent geometry, linking the buildings with an enclosed space which is polygonal in plan and undulating in elevation.

The office building facade consists of inner and outer walls separated by a series of 15-foot-wide terraces which step up the building along a diagonal. The diagonal offset in plan creates four additional corner offices on each floor, each with an adjoining terrace.

西地铁站

设计/竣工　1984/1986
伊利诺伊州，内珀维尔
威斯特敏斯特有限责任公司
232,000平方英尺
钢框架结构
彩色涂层组合式幕墙系统，反光灰色玻璃和珍珠光泽浅灰色花岗石火烧板

西地铁站是位于芝加哥都市西面的一个办公及科研园内的一栋10层高的出租用办公大楼。该项目包括以分期开发方式建设的两栋大楼，由一个被包围在其中的中庭结构将它们连接起来。该设计将一个抬高了的绿化地块用作整个建筑物的基座。

办公楼的立方体造型与列车月台区域的轴线之间成45°角。曲线型的中庭结构从建筑的对向面上悬垂下来，并成为一个独立的几何形体，将建筑物与一个平面为多边形、立面起伏变化的封闭空间连接起来。

办公大楼的立面由内层和外层墙体组成，两层墙体由15英尺宽的屋面平台隔开，平台沿立面的对角线层层上升。这种对角线上的退台在每一层楼上都形成了四个附加的转角办公室，每间办公室都拥有一个连在一起的平台。

1　Entrance with terraces
2　View from the south-west
3　Typical floor plan
4　Punched window within granite wall

1　入口及屋面平台
2　从西南向看建筑
3　标准层平面图
4　墙面开窗及花岗石墙面

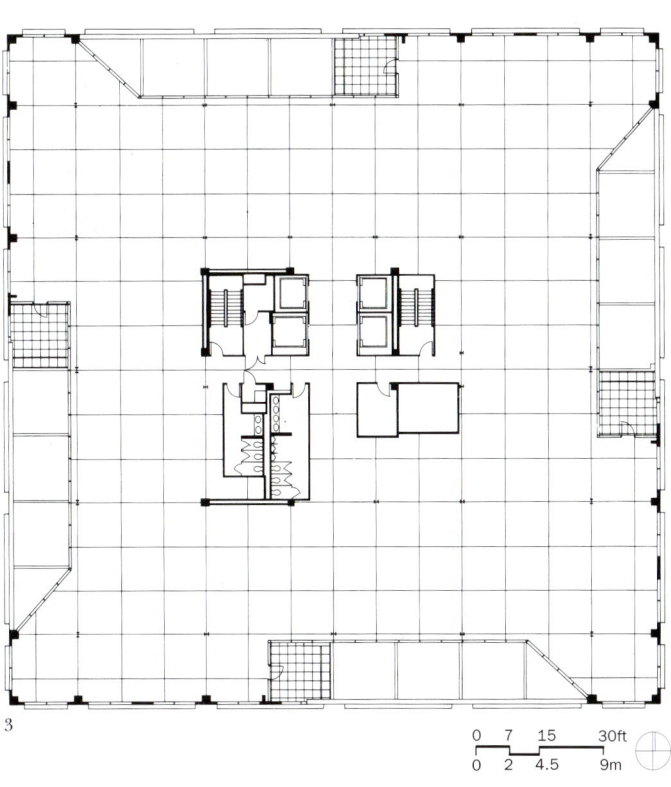

Metro West 117

South Ferry Plaza

Design 1986
New York, New York
Olympia & York Equity Corporation
1,200,000 square feet
Steel frame with metal deck
Aluminum and glass curtain wall

南渡口广场

设计　　　1986
纽约州，纽约市
奥林匹亚与约克证券公司
1,200,000平方英尺
钢框架结构及金属楼面
铝材和玻璃幕墙

Envisioned as a symbolic entry to New York City, the South Ferry Plaza Project transforms a confused traffic nexus into a dramatic sequence of public spaces that culminates in a 50-story tower marking the juncture of water and land transit.

The project is situated at the southern tip of Manhattan along the edge of the East River. Its public components include a new ferry terminal for the Staten Island Ferry, an observation plaza, and the Museum of Maritime History, all gathered beneath the welcoming arches of the commercial office tower. Monumental and maritime influences combine in the form of a lighthouse or harbor beacon that confidently marks Manhattan's prow.

The tower shaft rises out of a triangular base that gracefully transforms its three-sided plan into a hexagon. Great arches at the tower base channel gravity loads to three points straddling existing subway lines, Battery Park underpass, ferry slips and related terminal spaces, thus allowing virtually obstacle-free circulation space at grade.

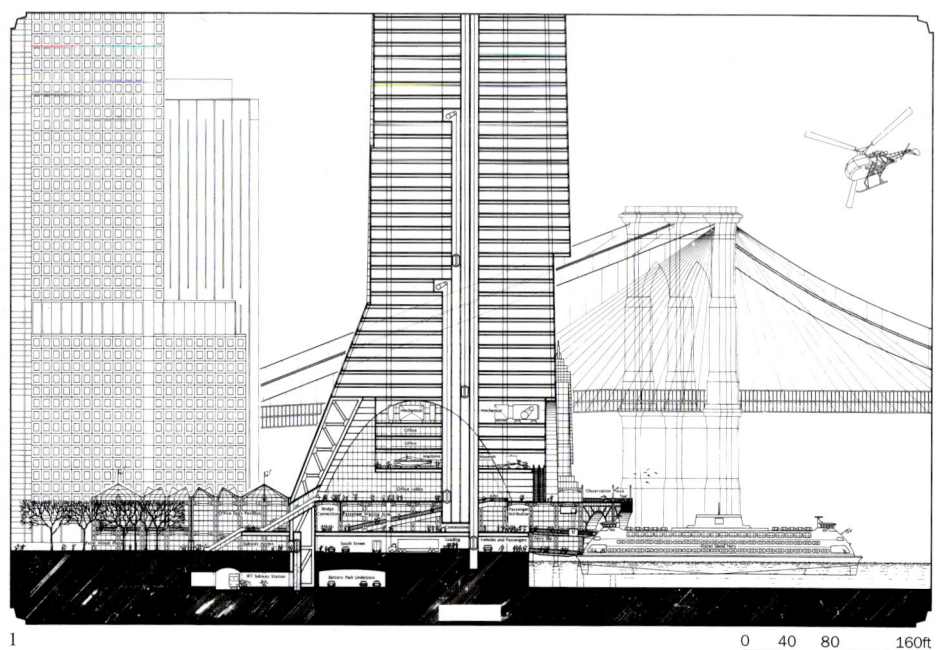

1

2

南渡口广场工程被设计为通往纽约市的一个象征性入口，它将混乱的交通节点改造成为一个充满戏剧性的公共空间序列，最终汇集在一个50层高的塔楼上，形成了水上和陆上运输的交汇处。

该项目位于沿东河边缘的曼哈顿岛的南端。其公共设施部分包括一个供斯德腾岛渡船使用的新渡船终点站、一个观光广场以及海洋历史博物馆，所有这些都聚集在商用办公塔楼形成的欢迎式拱门之下。纪念性和海洋性的影响被结合进了灯塔或港口信号塔的造型上，充满自信地标识出曼哈顿岛的最前端部分。

塔楼的主体从一个三角形的基座处升起来，以一种优雅的方式将由三条边所形成的平面转变成六边形。位于塔楼基座处的巨大拱门将大楼的重力荷载传递到三个支点上，这些支点跨越了现有的地铁线、巴特丽公园地下通道、轮渡码头港及相应的终点站空间，从而形成了在同一水平面上完全没有任何阻碍的流通空间。

1　剖面图显示出斯德腾岛渡船设施
2　城市景观

1　Section with Staten Island Ferry facilities
2　Cityscape

Columbus Circle

Design 1985
New York, New York
Trump Organization
2,100,000 square feet
Steel frame/metal deck
Glass and aluminum curtain wall

哥伦布环道

设计　　1985
纽约州，纽约市
王牌组织
2,100,000 平方英尺
钢结构，金属楼面
玻璃与铝材幕墙

The development site, located on Columbus Circle between 58th and 60th Streets, is currently occupied by the New York Coliseum. The project is envisaged as a multi-use complex of office, hotel and residential spaces.

The base structure contains retail space and new subway entrances. Office and residential entries are located at 58th Street, with additional office and hotel entries at 60th Street. Five floors of loft-type office space are also provided.

The tower is conceived as a segmented hollow octagonal tube 1,275 feet in height. A 230-foot-high slot is provided to reinforce the 59th Street view corridor. The tower is set back at skylobby level, 300 feet above grade. This level houses the main residential and hotel lobbies. The remainder of the tower terraces upwards in a spiralling configuration, commanding views of Central Park and the Hudson River.

该开发项目位于第58号大街和第60号大街之间的哥伦布环道上，现在，该场地被纽约大娱乐场所占据。该项目被设计成一个包括了办公、旅馆以及居住空间的多功能综合建筑物。

在建筑的裙房部分设置了商业零售空间和一些新的地铁入口。办公和住宅的入口位于第58号大街上，而在第60号大街上设置了另外的办公入口和旅馆入口。该项目中也提供了5层楼的大空间型的办公空间。

该塔楼被构思为一个成片段的空心八边形筒体，建筑的高度为1,275英尺。一个230英尺高的狭槽增强了第59号大街方向上的视线走廊。塔楼在距地面300英尺高的空中大堂平面上进行了退台处理，在这一层平面上设置了主要的住宅和旅馆大堂。塔楼的其他部分以一种螺旋式的构图向上逐渐退台，由此可以俯瞰中央公园和哈得孙河的美景。

1

1　从中央公园方向看建筑

1　View from Central Park

Bank of Southwest Tower

Design 1985
Houston, Texas
Century Development Corporation
Associate Architects: Lloyd Jones Brewer Associates
2,500,000 square feet
Composite steel and concrete structural system; eight perimeter "super columns" tied to concrete core with steel diagonals
Vertical aluminum tube mullions with reflective glass panels, stone spandrels with horizontal glass infill at corners

西南银行塔楼

设计　　1985
得克萨斯州，休斯敦
世纪开发有限公司
合作建筑设计单位：劳埃德·琼斯·布鲁尔联合事务所
2,500,000平方英尺
钢与混凝土混合结构体系，8根外围"超级柱子"由钢对角拉杆与混凝土核心拉结
垂直铝管竖框与反射玻璃板，
石材窗下墙饰面，转角处填充以水平玻璃

The design for the Bank of Southwest Tower juxtaposes the spirit and richness of past forms with present-day technique and materials. The obelisk, rotated on the site, is the basic formal generator of the design. The facades of the obelisk are framed by the major columns. The shift which occurs at these columns between the building's sloped and stepped facades reinforces the structure as load-bearing, and the nature of the skin as infill.

The surfaces of the building are treated in different combinations of granite, aluminum and glass, projecting an institutional and timeless character. The base is of granite to emphasize its strength and solidity, and uses classical forms in its detailing for the triangular arches and gables. The eight columns are faced with stone and merge with the stone gables at the top of the building.

西南银行塔楼的设计通过对当今先进技术及材料的应用来将其风貌与以往的建筑造型相融合。通过将方尖碑造型在基地上的旋转，使之成为本设计的基本形式的变形。方尖碑的立面上用大型的柱子来形成框架。在倾斜和退台的立面之间的柱子的平移强化了作为承重体系的结构和作为填充围护体系的外墙的特性。

建筑物的表面采用不同的花岗石、铝材及玻璃组合，以突出该建筑的公共性和永恒性的特点。建筑的基座部分采用花岗石，以强调其力量及坚固性，而三角形的拱门和山花则在细部处理上采用了古典形式。8根柱子以石材饰面，并在大楼的顶部与石材山花融为一体。

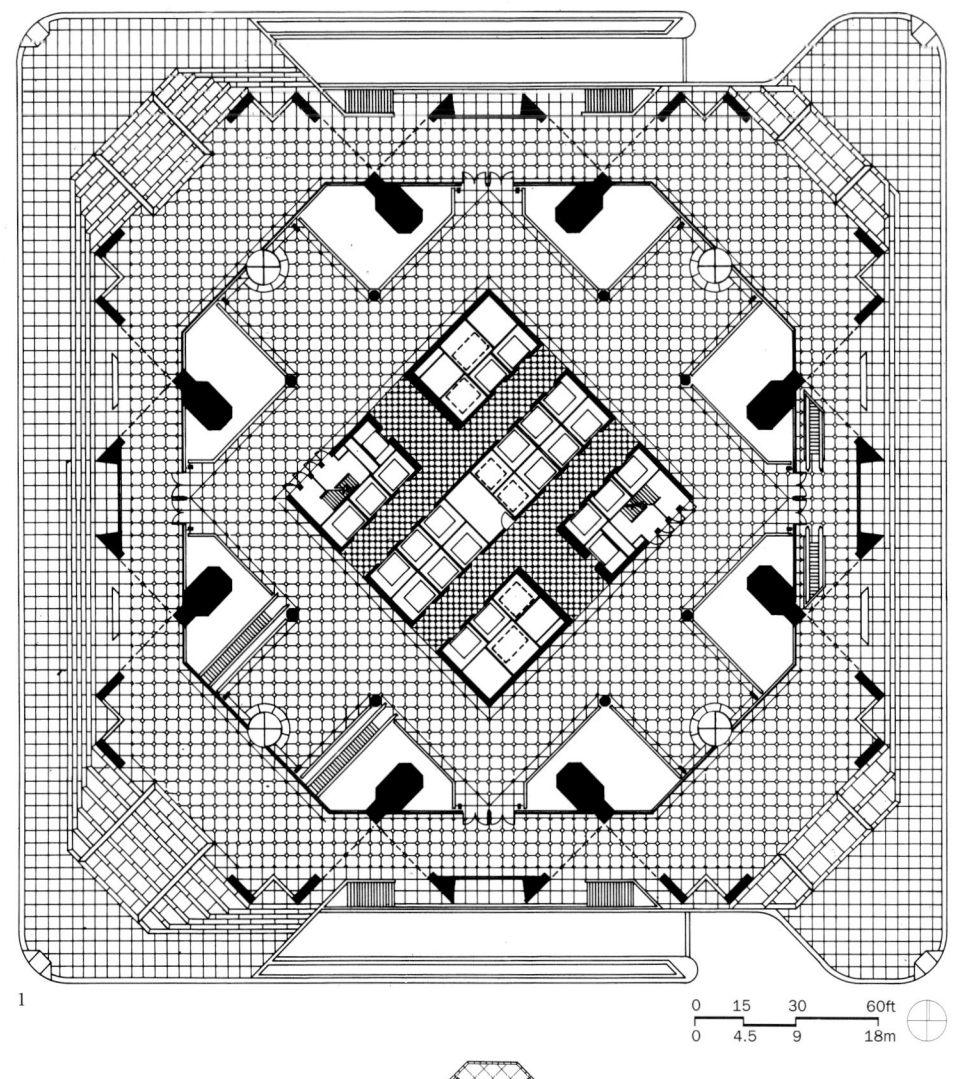

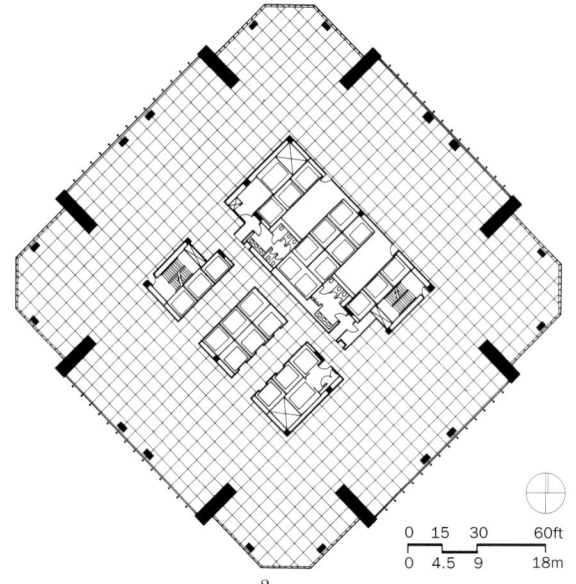

1　总平面图
2　标准层平面图
3　在城市天际线中旋转后的方尖碑

1　Site plan
2　Typical floor plan
3　Rotated obelisk on city skyline

11 Diagonal Street

Design/Completion 1981/1984
Johannesburg, South Africa
Anglo American Properties
Associate Architect: Louis Karol Architects
350,000 square feet
Concrete structure
Double wall skin, blue and silver reflective glass with red steel exposed frames supporting exterior curtain wall

斜向大街第 11 号

设计/竣工　　1981/1984
南非，约翰内斯堡
盎格鲁·美利坚房地产公司
合作建筑设计单位：路易斯·卡罗尔建筑师事务所
350,000 平方英尺
混凝土结构
双层外墙面，蓝色和银色反光玻璃与红色暴露式钢框架所支撑的外围幕墙

The site for this building is a full block in an area known as Newtown adjacent to the central business district of Johannesburg. The building takes the form of a diamond, both in reference to the diamond industry association of the developer's parent company, and to conform to the city's sloping height restrictions.

The building enclosure consists of a double skin. The inner wall is 50 per cent glass in a continuous strip, butt-jointed and silicone-sealed. This allows complete flexibility in locating interior partitions along the exterior wall. The space between the inner and outer walls provides an environmental buffer which is naturally ventilated. The exterior wall of reflective glass in aluminum frames acts as a sun shade, reducing solar gain.

该建筑的基地是在一个被称为牛顿区里的一个完整的街区地块，紧挨着约翰内斯堡市的中央商务区。该建筑采用钻石式的造型，既寓意着该开发商的母公司与钻石工业之间的联系，也满足了约翰内斯堡市关于斜向高度的规定。

该建筑的围合体由双层外墙构成。内侧墙体部分使用了 50% 的玻璃，与呈连续的条带状的墙体平接并用硅树脂进行密封。这种做法可使在沿外部墙体设置室内隔墙时具有完全的灵活性。在内外墙之间的空间形成了可以自然通风的环境缓冲带。外墙采用铝框嵌反光玻璃，起着遮阳的功能，以降低室内温度。

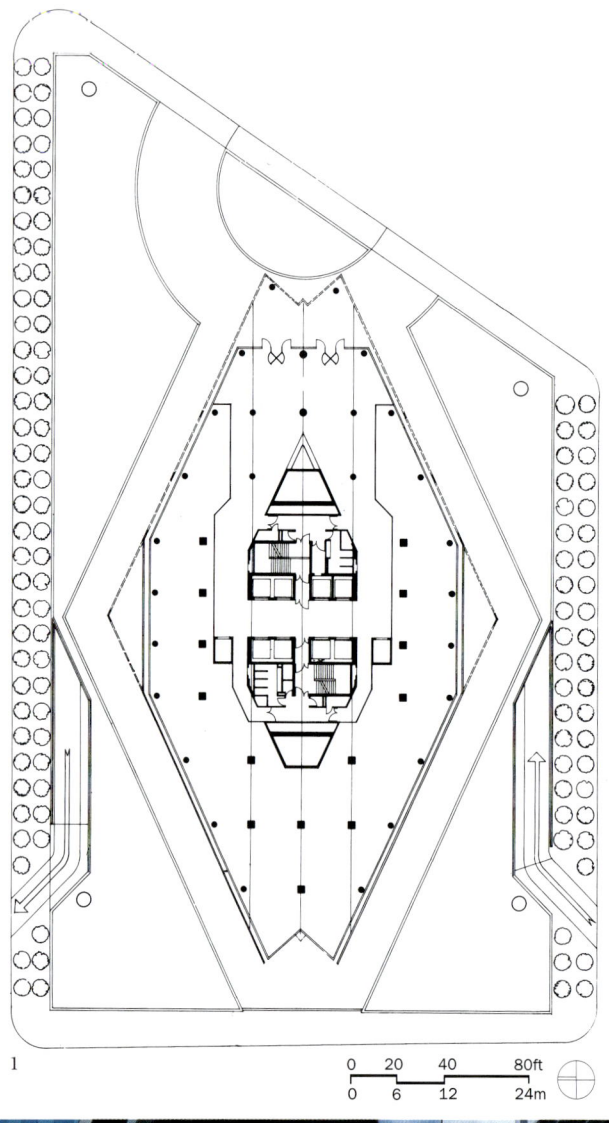

1　总平面图
2　入口部分
3　城市景观
4　门厅内景
5　正立面

1　Site plan
2　Entrance
3　Cityscape
4　Lobby
5　Front elevation

3

4

5

11 Diagonal Street

Chicago Board of Trade Addition

Design/Completion 1978/1982
Chicago, Illinois
Chicago Board of Trade Joint Venture
Architects: Shaw and Associates; Swanke Hayden Connell Architects
584,000 square feet
Steel frame supporting two-story-high trusses over trading floor; glass roof supported on steel trusses
Aluminum curtain wall grid module, silver reflective vision/spandrel glass in combination with dark gray vision/spandrel glass and limestone veneer

芝加哥同业公会扩建工程

设计　　1978/1982
伊利诺伊州，芝加哥市
芝加哥同业公会联合企业
合作建筑设计单位：肖及合伙人事务所；史瓦克·海登·康内尔建筑师事务所
584,000平方英尺
由钢框架支撑位于交易楼层上的两层楼高的桁架，由钢桁架支撑玻璃屋面
铝制幕墙网格构件，银色反射观景玻璃和窗下墙玻璃与深灰色观景和窗下墙玻璃及石灰岩饰面板的组合

The building provides 584,000 gross square feet for the Chicago Board of Trade, including a new trading floor and support spaces as well as office space for exchange members and staff. The addition is designed to function with the existing building as one unit.

The addition responds functionally and formally to the existing Art Deco landmark structure. The first 12 floors, housing the trading floor and support activities, correspond to similar spaces in the existing structure. Above this level, office floors are designed as U-shaped spaces around a central atrium which adjoins the existing structure. To provide the required trading floor area, the building projects 20 feet beyond the existing structure, creating a covered pedestrian arcade at street level.

A glass skin wraps the highly articulated planes of the wall and the roof.

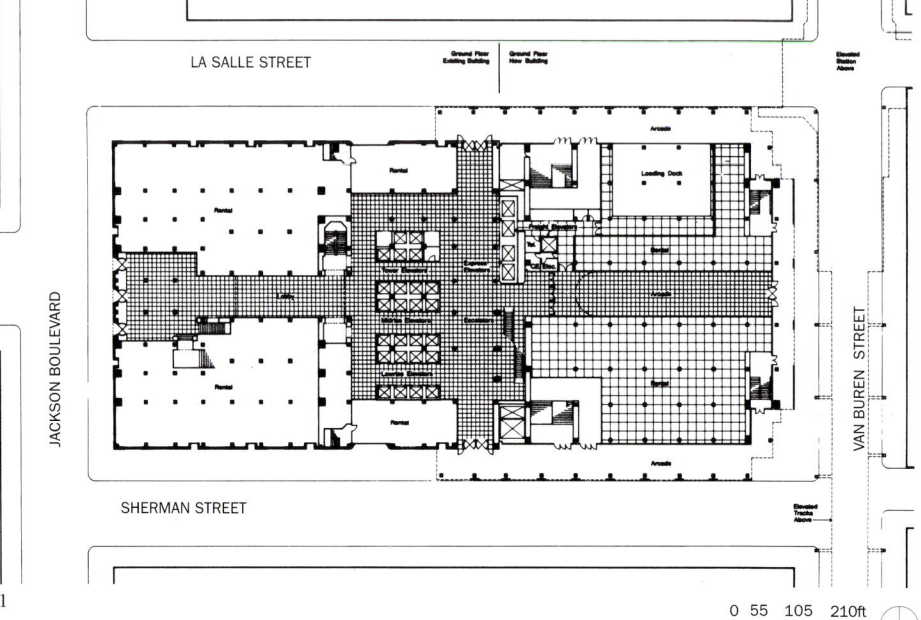

该建筑为芝加哥同业公会提供了584,000平方英尺的总建筑面积，包括一个新的交易楼层和辅助空间，以及供交易成员和工作人员使用的办公空间。扩建工程部分被设计来与原有建筑作为一个单元来使用。

扩建工程部分在功能与形式上与现有的艺术装饰派风格的地标性建筑形成呼应。建筑下部的12层中设置了交易楼层和辅助活动空间，与原有大楼中相近似的空间相一致。在该标高以上的办公楼层被设计成U形空间，围绕着一个与原有大楼相连接的中庭来布置。为了达到对交易区建筑面积的要求，该建筑从原有建筑上延伸出了20英尺，在街道平面上形成一个有顶的人行廊道。

该建筑使用一层玻璃外皮将极具表现力的墙面和屋面整个包裹起来。

1　底层平面图
2　从西南向看建筑
3　在第十二层上的中庭

1　Ground-floor plan
2　View from the south-west
3　Twelfth floor atrium

One South Wacker

Design/Completion 1979/1982
Chicago, Illinois
Metropolitan Structures/Harvey Walken & Co.
1,280,000 square feet
Reinforced concrete structure
Aluminum and glass curtain wall; terrazzo floors with marble and glass lobby

南瓦克一号

设计／竣工　1979/1982
伊利诺伊州，芝加哥
大都会建筑公司及哈维·沃尔肯公司
1,280,000平方英尺
钢筋混凝土结构
铝和玻璃幕墙，水磨石楼面，门厅部分为大理石与玻璃

This is essentially an adaption of the typical office tower of the 1920s to today's standards and requirements. The concept of the building is based on providing very large floor areas, ranging from 25,000 to 38,000 gross square feet. The building is set back twice, giving three typical floor areas. Three-story atriums located above and below the setbacks create U-shaped floor areas and thus reduce the depth of the lease-span and increase the perimeter and daylight exposure.

At the ground floor the atrium extends into a multi-level commercial galleria which serves as a civilized public amenity, providing a through-way between the Wacker and Madison entrances to the building. Commercial functions are kept to the exterior, in keeping with the character of the adjacent streetscape.

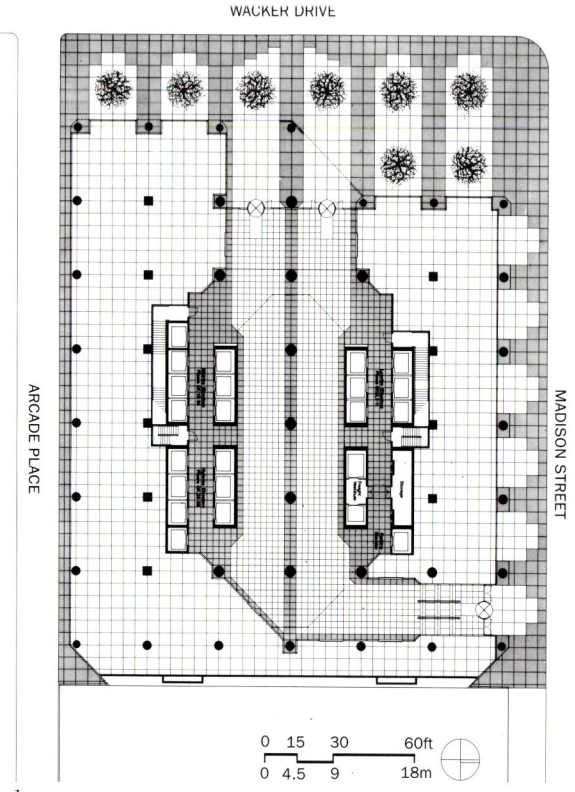

该项目实际上是对20世纪20年代的典型办公塔楼进行的一种改进设计，以使其满足今天的建筑标准和要求。该项目在设计构思上是要提供相当大的楼层建筑面积，从25,000到38,000平方英尺。建筑有两层退台处理，形成了三种标准层建筑面积。在退台处的上面和下面有3层高的中庭，形成U形的楼层平面，从而减少了可供出租的空间的跨度，也增加了建筑物周边长度和采光面。

在底层上的中庭一直延伸到一个多层的商业长廊部分，该长廊被用于公共休闲，并在瓦克大街和麦迪逊大街方向的大楼入口之间形成了一条便捷的通道。商业功能部分均朝向建筑外侧，以保持邻近的街区景观的特色。

1 底层平面图
2 入口门厅
3 从瓦克大街上向南看建筑
4 瓦克大街立面夜景
5 从芝加哥河上看到的城市环境
6 建筑入口

1 Ground-floor plan
2 Lobby
3 Looking south on Wacker Drive
4 Wacker Drive elevation at night
5 City context with Chicago River
6 Entrance

3

4

5

6

One South Wacker 127

Xerox Centre

Design/Completion 1978/1980
Chicago, Illinois
Romanek-Golub
880,000 square feet
Reinforced concrete cast-in-place
Unitized aluminum panel and insulated glass system

施乐中心

设计／竣工　1978/1980
伊利诺伊州，芝加哥市
罗曼内克·格鲁伯公司
880,000 平方英尺
现浇钢筋混凝土
组装式铝制镶板与隔热玻璃系统

The building's typology is that of an "infill building on a corner". Like the existing buildings in the block, it extends to the property line along Monroe Street and is set back on Dearborn, continuing the landscaped promenade established by the First National Bank. It is also set back at Marble Court to provide light and views on the lower levels. The building curves at the corner to make the transition between its two sides and symbolically suggest its treatment as a single front. At the ground floor the curved wall continues to form recessed entries; at the roof it makes the transition to the mechanical penthouse.

Above the height of the existing structures on Monroe Street, the building slopes to assume its tower configuration. This provides a strong continuity between old and new, and is a break with the concept of a "freestanding tower" on a plaza.

该项目在类型上属于一个"在角落上插入的建筑"。与在该街区上的原有建筑群一样，该建筑延伸至沿蒙洛大街的建筑红线处，在迪尔伯恩大街方向上后退缩进，使从第一国家银行处开始形成的观景步行道得以延续。该建筑也在马贝尔庭院处后退红线，以便为较低楼层提供采光和观景的条件。大楼在这一角落处成曲线形，以形成建筑的两个面之间的过渡，并象征性地暗示着将其作为单一立面来进行处理。在底层平面上，弯曲的墙面延伸出来，形成凹入式的入口，而其屋面部分则自然过渡到设备用房的造型上。

该建筑在高出位于蒙洛大街上的原有建筑物的部分形成倾斜状，以表现出其塔楼式的造型。这种设计形成了在新旧建筑之间的一种强烈的延续性，突破了一般在广场上建造一个"独立式塔楼"的构思。

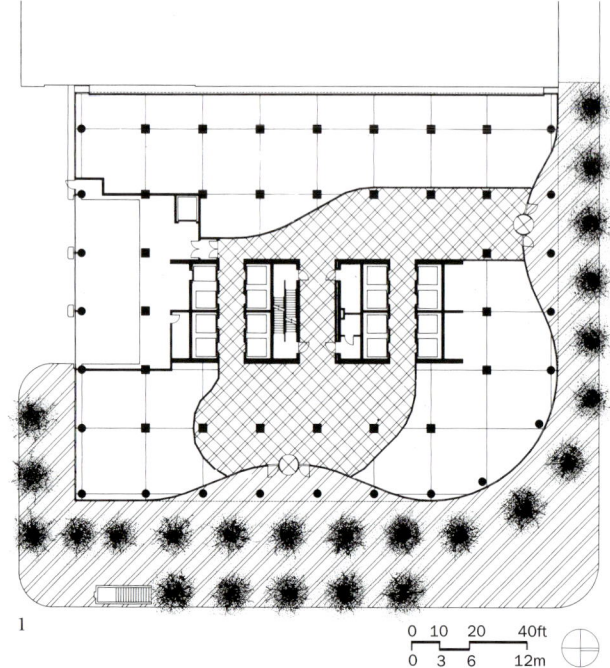

1　总平面图
2　在转角处的立面过渡处理
3　从东北向看建筑全景

1　Site plan
2　Transition of facades at corner
3　Overall view from the north-east

作品精选

城市街区建筑

- 132 库-达姆70号
- 136 雅典娜办公大楼
- 142 查理曼大帝大厦
- 143 第一信托人寿保险公司
- 144 斯特拉劳尔广场35号
- 148 柏林索尼中心
- 150 维多利亚保险公司
- 152 库-达姆119号
- 154 曼海姆人寿保险公司总部大楼
- 158 奥古斯丁霍夫——纽伦堡
- 160 欧罗巴住宅
- 161 汉莎航空公司总部大楼
- 162 哈罗德·华盛顿图书馆中心设计竞赛
- 163 弗斯特与百老汇洛杉矶市政中心
- 164 伊利诺伊州立中心
- 168 桑德·莫拉罕集团公司总部大楼
- 170 费斯特·索尔斯中心

Selected and Current Works

Urban Block Buildings

- 132 Ku-Damm 70
- 136 Pallas Office Building
- 142 Charlemagne
- 143 Principal Mutual Life Insurance Company
- 144 Stralauer Platz 35
- 148 Sony Center, Berlin
- 150 Victoria Versicherungsgesellshaft
- 152 Ku-Damm 119
- 154 Mannheimer Lebensversicherung Headquarters
- 158 Augustinerhof-Nuernberg
- 160 Europa-Haus
- 161 Lufthansa Corporate Headquarters
- 162 Harold Washington Library Center Competition
- 163 First & Broadway Los Angeles Civic Center
- 164 State of Illinois Center
- 168 Shand Morahan Corporate Headquarters
- 170 First Source Center

Ku-Damm 70

Design/Completion 1988/1994
Berlin, Germany
Euwo Unternehmensgruppe
12,000 square feet
Steel floors hung with rods from steel trusses supported by a concrete core
Curtain wall of fritted glass, aluminum mullions and aluminum panels

库－达姆70号

设计/竣工　　1988/1994
德国，柏林市
尤沃企业集团
12,000平方英尺
用拉杆悬挂于钢制桁架上的钢制楼板，桁架由一个混凝土核心筒来支撑
热熔玻璃、铝制竖框和铝板所组成的幕墙

This is the first building of a series on Ku-Damm. It is an urban repair project, covering the end of a building exposed through insensitive and destructive street planning in the 1950s. In its attitude the building is an urban intervention against the mediocrity of the new city along the Ku-Damm, evoking nostalgia for the lost Berlin tradition.

Due to an available site width of only 10 feet, the building was allowed to be cantilevered over the sidewalk on Ku-Damm and Lewishamstrasse, and to exceed the height of the adjacent building. This makes the building a marker on Ku-Damm, an image reinforced by its knife-like plan configuration and its steel mast with sign.

The steel floors are hung from the 10-foot-wide concrete service core. Cladding is of an all-glass skin, accentuated by the hangers and alternating floors. The glass is fritted to create different transparencies in response to the view conditions from the inside, resulting in a decorative pattern within this strong construct.

这是位于库－达姆大街上的一系列建筑群中的第一栋大楼。这是一项城市修复工程，要将一栋大楼暴露在一条于20世纪50年代规划的道路上的山墙进行整修。该建筑实际上是一项城市改造工程，以改变沿库－达姆大街方向上的新建市区所表现出的平庸无趣的面貌，同时唤起公众对已经消失了的柏林传统建筑风格的怀旧之情。

由于可用的基地宽度仅为10英尺，该建筑被允许在库－达姆大街及列维史曼大街的人行道上挑出来，并且可以超过毗邻建筑物的高度。这使得该建筑成为库－达姆大街上的标志，也形成了其刀形的平面构成关系和带有标志物的钢制桅杆。

钢制的楼板从10英尺宽的混凝土设备核心筒处悬挂下来。建筑的外墙是全玻璃的面层，重点处理悬挂件和交错构图的楼层。玻璃经过了热熔处理，从而呈现出不同的透明度，与在建筑物内部观景的不同条件相呼应，也在这个坚固的建筑形象上形成了一种装饰性的图案。

1

1	Ku-Damm elevation	1	库－达姆大街方向的立面
2	Concept sketches	2	构思草图
3	Plans and section	3	平面图和剖面图
4	Knife-like plan with mast	4	刀形的平面和桅杆
5	Steel mast detail/sign creating a symbol on Ku-Damm	5	位于钢桅库－达姆大街方向的桅杆细部和标志物

2

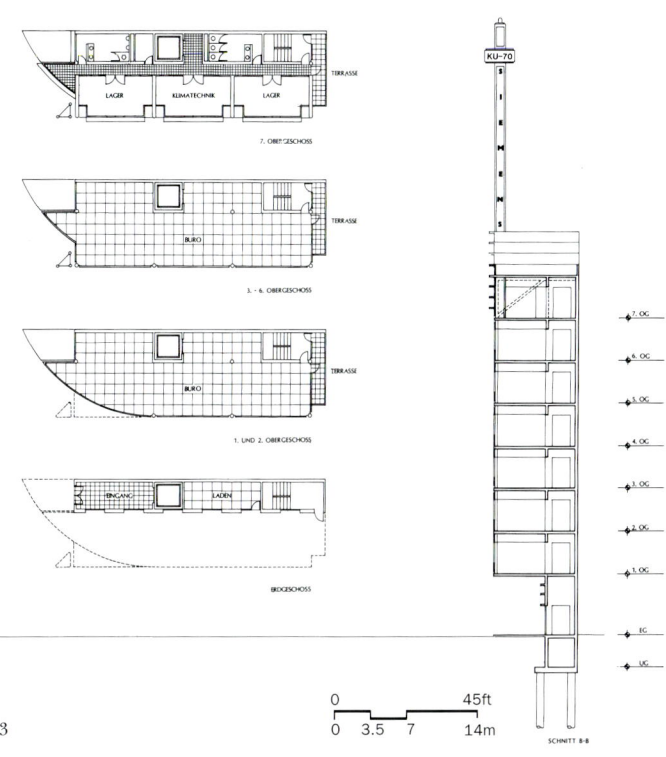

3

4

5

Ku-Damm 70 133

6–9 Curtain wall details with fritted patterned glass
10 Knife-like plan with steel mast

6-9 热熔玻璃的幕墙细部
10 刀形的平面和钢桅杆

6

7

8

9

Pallas Office Building

Design/Completion 1989/1994
Stuttgart, Germany
IGEPA Industrie und Gewerbepark Bautraeger
600,000 square feet
Concrete structure
Curtain wall of light and dark granite, glass and aluminum

The site for this project is located in a suburb of Stuttgart which is in transition from light industry to office and service uses. The existing cityscape presents a random collection of buildings based primarily on industrial prerogatives.

The building mass uses the clearly articulated streetwalls of the traditional European city block to establish a strong exterior presence and reveal a new interior urban space. The building extends to the boundary on all sides, defining a solid block with rounded corners. A primary gate facing the main train station provides automobile access to the three main lobbies. Secondary gates at the corners of the courtyard transform the simple wall masses into tower elements.

In the courtyard stand three glass wedges set into black granite walls, defining the building's main entrances. At the base of the wedges are high stone arcades leading to the main lobbies.

该项目的基地位于斯图加特的一个郊区，该地区正从轻工业制造转向办公和服务功能。现有的城市景观表现为一种主要基于工业特征上的杂乱无章的建筑群的布置。

建筑在体量上采用传统的欧洲城市街区建筑那种清晰的连续式沿街墙处理方式，从而使建筑物形成了一种强烈的外部形象，也展现出一种新的内部城市空间。该大楼在各个方向上向边界处扩展，从而限定了一个带有圆形转角处理的实体街区地块。一个主要的面向火车站的入口大门形成了通往三个主要大厅的车流通道。在庭院转角处的次要大门将单一的墙式体量转变成了塔楼式的元素。

在庭院内部耸立着三个嵌入黑色花岗石墙体的玻璃楔形体量，标志着建筑物的几个主要入口。在楔子的基座处是高敞的石制廊道，一直通向主要的大厅。

1 总平面图和底层平面图
2 入口大门
3 城市街区建筑
4 塔楼式的元素
5 花岗石墙面和廊道

1 Site/ground-floor plan
2 Entrance gate
3 City block
4 Tower element
5 Granite wall and arcade

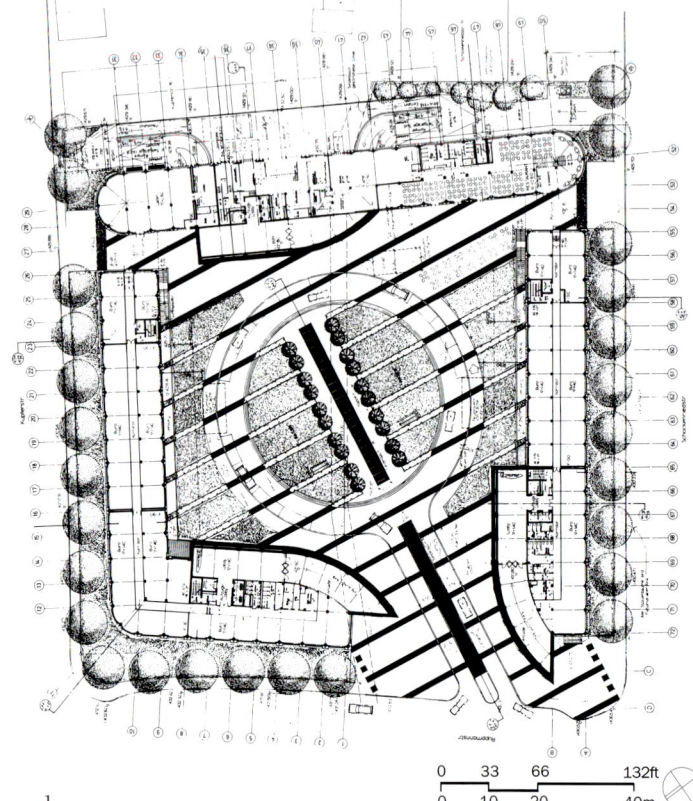

Pallas Office Building 137

6

7

8

6	Courtyard	6	中央庭院
7	Glass wedges	7	玻璃楔形体量
8	Entrance drive	8	入口处的车道
9	Secondary gate	9	次要入口
10	Punched granite wall	10	开窗的花岗石墙面
11	Glass wall	11	玻璃墙面
12	Entrance gate	12	入口大门

9

10

11

12

13	Arcade	13	廊道
14	Lobby arcade	14	大厅前的廊道
15	Elevator corridor	15	电梯走廊
16	Lobby	16	大厅

13

14

15

16

Pallas Office Building 141

Charlemagne

Design/Completion 1994/1997
Brussels, Belgium
Etudes et Investissements Immobiliers, SA/NV
Associate Architect: ARC, SA Architectes
614,000 square feet
Steel structure, glass bay point and screen wall additions
Clear and patterned fritted glass with printed aluminum mullions; stone-clad base with button-glazed lobby front

查理曼大帝大厦

设计/竣工　1994/1997
比利时，布鲁塞尔
埃突德不动产投资公司，SA/NV
合作建筑设计单位：ARC 建筑设计事务所，SA 建筑设计事务所
614,000 平方英尺
钢结构，大玻璃开间处理及附加幕墙
透明和压花热熔玻璃与涂色铝材坚框，
基座处用石材贴面，门厅部分立面点缀玻璃墙面

The Charlemagne project is an addition and remodeling of the existing Charlemagne building at Rue de la Loi 170, Brussels. Constructed in 1968, this 15-story Y-shaped building is situated on a three-story parking base which is exposed at the Rue du Taciturne due to the slope of the site. The existing building contains 431,733 square feet of office space above grade.

This project modifies the existing building in several ways. A 5.5-foot addition wraps the existing Y, forming a 71-foot tri-span office plan. The ends of the east and west wings are extended to the site boundary. Two curved winter gardens fill the void between the three wings of the Y. The center wing is cut back to unify the curved shape of the two winter gardens. The top of the parking base is transformed into a raised urban garden terrace accessible from the two winter gardens.

查理曼大帝大厦位于布鲁塞尔的拉·罗伊大街170号，该项目主要是对原有的查理曼大帝大楼进行扩建和改造。现有的大楼修建于1968年，平面为"Y"字形，大楼的基座部分是一个3层高的停车库，上面还有15层。由于该大楼所在的基地为一个斜坡，因此，该大楼的基座在临塔西特纳大街的部分才能够完全敞开。现有建筑在地平面以上有13,733平方英尺的办公区建筑面积。

该项目采用了不同的措施来对原有大楼进行改造。用一个5.5英尺宽的附加体将原有的"Y"字型大楼包裹起来，形成一个71英尺深的三跨的办公区平面。东西两翼的山墙被延伸到基地的边界处。在"Y"字型大楼的三个侧翼之间的空白处设置了两个曲面的温室花园。而中央的一翼则被拆掉一部分，从而使两个温室花园的曲线外观统一起来，停车库基座的顶部被改造成一个在屋面上的城市花园平台，人们可以从两个温室花园到达这个平台。

1　曲面的温室花园立面
2　拉·罗伊大街方向上的遮阳幕墙

1　Curved winter garden facade
2　Sun-shade screen wall on Rue de la Loi

Principal Mutual Life Insurance Company

Design/Completion 1993/1995
Des Moines, Iowa
Principal Mutual Life Insurance Company
518,000 square feet
Structural steel frame with composite deck
Clear and patterned spandrel glass, fluoropolymer-coated aluminum panels, limestone cladding

第一信托人寿保险公司

设计/竣工 1993/1995
艾奥瓦州，得梅因市
第一信托人寿保险公司
518,000平方英尺
钢框架结构与混合结构楼板
透明和压花玻璃窗下墙玻璃，含氟聚合物涂层铝材镶板饰面，石灰石贴面

The project program called for 456,650 gross square feet of new office space to be developed on the site in downtown Des Moines. The plan creates a public plaza between the new building, at the north end of the site, and Corporate Square to the south, linking the two with a public skywalk.

The building is developed on a broad footprint at the north end of the site. The large floor plates enable the mass of the building to be reduced in height and conform to the scale of the immediate context. The plan of the building can be interpreted as a long, linear element which has been "folded up" to fit the site, and simultaneously to define the street spaces to the north, east, and west.

In one interpretation, the building becomes a "wall" determining the north edge of the plaza; in another sense, the building becomes a giant "gateway" linking areas to the north and south.

该项目的计划上要求在这个位于得梅因市中心的基地上开发出总建筑面积达456,650平方英尺的新办公空间。在基地的北端处，该方案在新建筑物之间设计了一个公共露天广场，并且用一个公共人行天桥将它与位于该基地南面的联合广场连接起来。

该大楼项目位于基地北端的一个宽阔的地块上。巨大的楼板使得大楼体量在高度上能够得以弱化，并与四周紧邻的文脉环境在尺度上取得一致。该大楼的平面可以被看成是一个长长的线形建筑元素，该元素被"折叠"起来以适应建筑基地的特征，同时也以之限定了在北向、东向和西向的街道空间。

在某种程度上可将该建筑物看成是一面墙体，它确定了露天广场的北侧边缘。而从另一种角度来看，该建筑变成了一个巨大的入口，将其南北两侧的区域联系在一起。

1

2

1 Folded linear footprint of building with wall and gate
2 Public plaza connecting the new building with existing corporate offices

1 折叠式的线形建筑及由之形成的墙体和大门
2 公共露天广场将新建筑与原有的公司办公大楼连接起来

Stralauer Platz 35

Design/Completion 1993/1997
Berlin, Germany
OPUS Entwicklungsgesellschaft
468,000 square feet
Placed concrete
White granite with gray granite base, patterned fritted glass cable net wall structures, coated steel and aluminum sun shades and Krone

斯特拉劳尔广场 35 号

设计/竣工　1993/1997
德国，柏林市
OPUS 开发公司
468,000 平方英尺
现浇混凝土
白色花岗石外墙与灰色花岗石基座，压花热熔玻璃与钢索网墙体结构，涂层钢和铝材遮阳板及顶部构架

Occupying a site formerly bisected by the Berlin Wall, the building is to be read as a permeable form rather than as a volume constructed of edges.

The high plaza is framed by two buildings, a moderate five-story structure and a stepped 15-story building addressing the park and corner. The open plaza extends the Stralauer Platz as the main entry to the building complex. A river promenade is created by lowering the grade along the constant-level riverfront. As future development continues, a wide pedestrian path will extend along the riverfront park.

The Spree stair connects the two levels. Incorporating both steps and landscaped terracing, the stair will become the type of informal gathering place which characterizes the habitable city. The winter garden, a high vertical space, terminates the expanse of the parklands to the east. The enclosed area mitigates the solar exposure on this face of the project and provides an extended comfort zone to cafe functions at the park's edge.

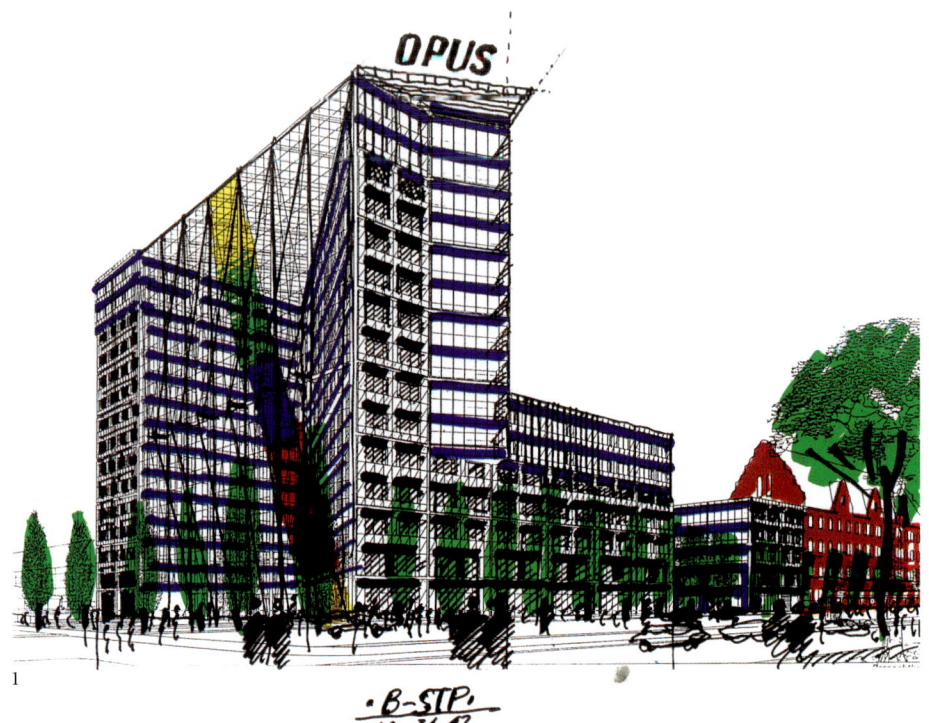

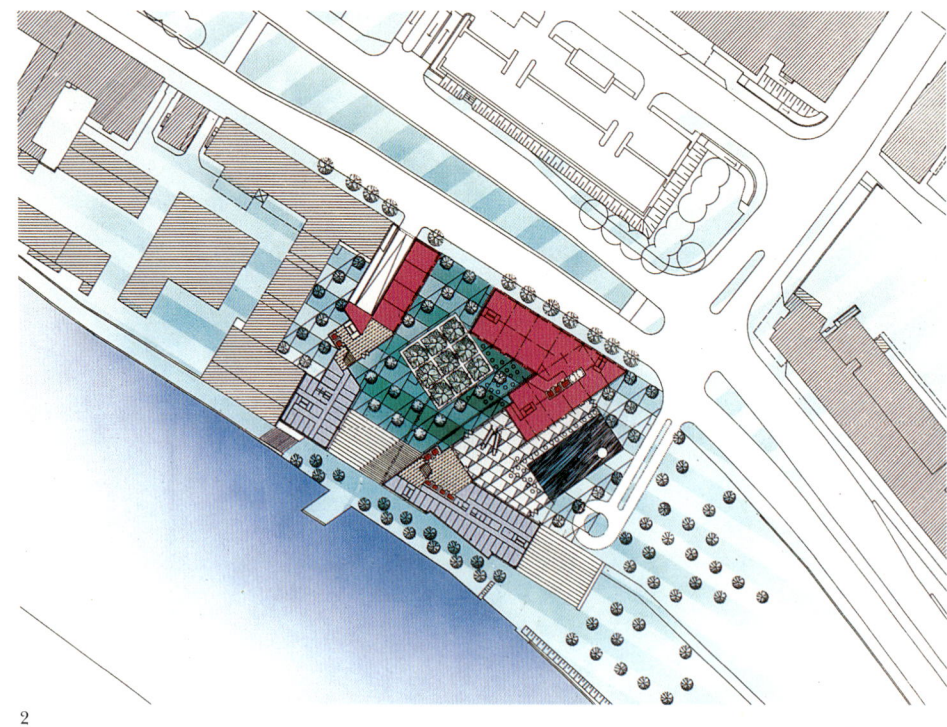

该建筑的基地原来被柏林墙分成两部分，因此，在建筑设计上将它处理成一个通透的造型，而不是由各个边所构成的一个巨大的体量。

庄重的露天广场由两栋大楼围合而成。一栋是体量适宜的5层建筑物，而另一栋做退台处理的15层建筑物强化了公园和转角处的处理。这个开放式的露天广场延伸了斯特拉劳尔广场，并且将它作为通往该综合建筑群的主要入口。通过降低沿着具有稳定水面高度的滨河地区的地坪标高而形成了一条滨河林阴步行道。随着后继的开发项目建设，将会出现一条沿着滨河公园延伸的宽阔的人行通道。

斯比里楼梯将这两个标高联系起来。该楼梯在设计上结合了踏步和绿化台，成为非正式的聚会场所，表现出这个适于居住的城市的特色。温室花园是一个有着高耸的垂直式空间，它结束了在其东面的公共绿地的扩张延伸。围合起来的区域减少了该项目在这个立面上的直接日照，并且形成了一个延伸出来的休闲区，成为在公园的边缘处的咖啡座功能的延续。

1	Concept sketch	1	构思草图
2	Site plan	2	总平面图
3	Winter garden	3	温室花园
4	Context	4	文脉环境

3
4

Stralauer Platz 35 145

5

6

7

5　Stepped 15-story tower
6　River promenade
7　Pedestrian path
8　Winter garden
9　River view

5　退台式的15层高的塔楼
6　滨河林阴步行道
7　人行通道
8　温室花园
9　沿河景观

Sony Center, Berlin

Design/Completion 1993/1999
Berlin, Germany
Sony Corporation
1,670,000 square feet
Reinforced concrete and steel frame
Naturally ventilated glass and aluminum curtain wall

柏林索尼中心

设计/竣工　　　1993/1999
德国，柏林市
索尼公司
1,670,000 平方英尺
钢筋混凝土与钢框架结构
自然通风的玻璃和铝材幕墙

The program for the Sony Center required spectacular office headquarters, a cinema and cultural complex, retail, a hotel, apartments, and parking.

The buildings are shaped to respond to the urban situations at the perimeter, such as the Potsdamer Platz, the boulevard of the Potsdamer Strasse, the Kemper Platz and the Bellevue-Park. Materials used on the facades are stone, metal and glass in different combinations and configurations which respond to each building's use and form.

The attic floors of the Sony headquarters frame a giant inward-sloping "show window", overlayed with a monumental order of columns, providing views into the atrium and exhibition space. At the Filmhaus a "screen" projects from its gridded and framed glass facade. The larger-scale elements of "show window" and "screen" respond to the faster movements along the Entlastungs Strasse and to the solitary buildings of the Kulturforum.

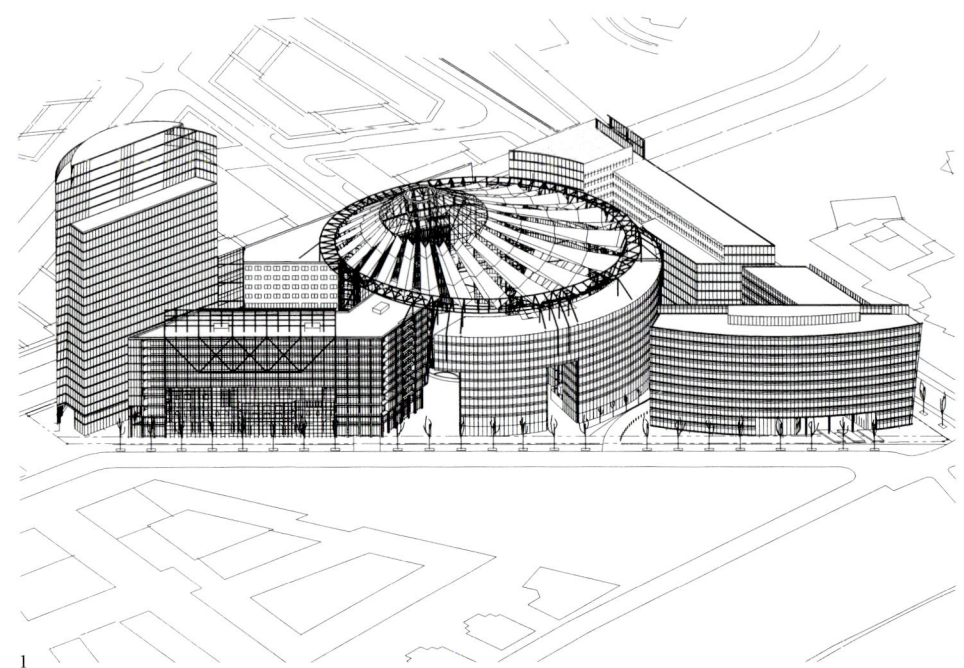
1

索尼中心的项目计划要求设计出形象壮观的办公总部大楼、一个电影院和文化演出综合建筑、商业零售区、一座酒店、部分公寓楼以及配套的停车场。建筑物在造型上尽力与四周的城市环境形成呼应，例如波茨坦广场、波茨坦大街上的林阴大道、坎帕尔广场以及贝勒乌公园。在建筑立面上使用的材料包括石材、金属和玻璃，应用了不同的组合方式和构图形式，使之与每栋建筑的用途和造型形成呼应关系。

索尼公司总部的阁楼层形成了一个巨大的向内倾斜的"橱窗"，在其外面饰以纪念性的柱列形象，可以在其中看到中庭及展览空间的美景。在电影院部分，一个"银幕"从建筑的格栅和框架式的玻璃立面上悬挑出来。"橱窗"和"银幕"这些尺度较大的建筑元素与沿着交通支路产生的高速运动和该文化广场上那些孤立的建筑物形成呼应关系。

2

3

1 Axonometric	1 轴测图
2 Bellevue Strasse elevation	2 沿贝勒乌大街的建筑立面
3 Entlastungs Strasse/Potsdamer Strasse	3 沿交通支路与波茨坦大街的建筑立面
4–7 Concept sketches	4–7 构思草图

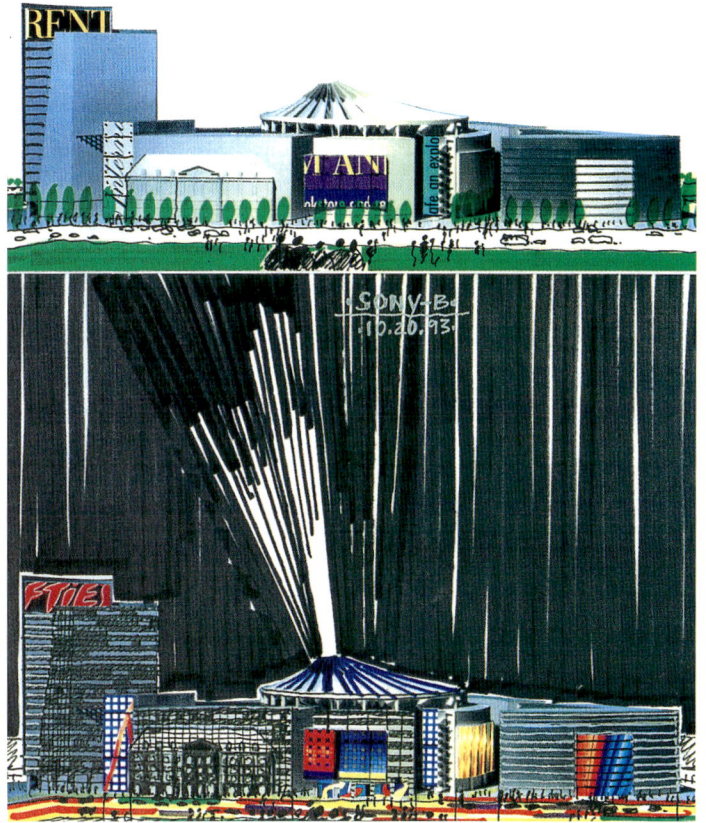

4

5

6

7

Sony Center, Berlin 149

Victoria Versicherungsgesellschaft

Design/Completion 1992/1997
Berlin, Germany
Victoria Versicherungsgesellschaft
823,400 square feet
Placed concrete, steel and aluminum, and glass roofs and canopies
Structural glazing system with clear and patterned fritted vision glass and spandrel glass

维多利亚保险公司

设计／竣工　　1992／1997
德国，柏林市
维多利亚保险公司
823,400平方英尺
现浇混凝土，钢材及铝制材料，玻璃屋面及雨篷
安装玻璃的结构体系及透明与压花热熔观景玻璃
和层间玻璃饰面

Based on a partially realized plan of 1957, the Victoria Project concludes the redevelopment of Berlin at the corner of the Kurfuerstendamm and Joachimstalerstrasse. Befitting the status of the Kurfuerstendamm as one of Berlin's pre-eminent streets, the new Victoria building emphasizes the defining eight-story high cornice at the streetwall. Except for the minor segment of retail development to be removed, the existing buildings will be upgraded. The existing eight-story Victoria building will be provided with a new, more accessible and gracious entrance.

The project deviates from the originally planned scheme by including over 100 residential units. The apartments are located away from the busy main streets at the north-west quadrant of the site. Retail functions will be provided at all newly constructed grade-level areas and some second-level areas. Existing retail areas will be upgraded.

维多利亚保险公司大楼是以按照1957年设计的方案而部分完工的建筑为基础的，该项目被作为柏林再开发计划的结束部分，位于库尔夫斯腾达姆大街和乔希姆斯达勒大街转角处。为了将库尔夫斯腾达姆大街建设成为柏林市最杰出的大街之一，新的维多利亚保险公司大楼强化了在沿街墙面上的具有决定性构图作用的八层处的挑檐板。除了将原有建筑上一些小块的商业零售区拆除外，原有建筑部分将被予以改造。原有的8层高的维多利亚保险公司大楼将增建一个新的、进出来往更为方便、造型优美的入口。

该项目与原来的规划布局有一些错位，新项目中包括了100多个住宅单元。公寓楼部分位于远离繁忙的主街道，将它布置在该基地的西北方向上。在新建部分的整个首层平面和二层平面的一部分区域设置了商业零售功能，而原有的商业零售区将被改造更新。

1

2

3

1 City context	1 城市文脉及周边环境
2 Passageway from Kantsrasse	2 从坎特大街方向延伸过来的通道
3 Kurfuerstendamm/Joachimstalerstrasse	3 从库尔夫斯腾达姆大街与乔希姆斯达勒大街方向看建筑
4 Ground-floor/site plan	4 首层平面图与总平面图
5 Rendering from Kurfuerstendamm/Joachimstalerstrasse	5 库尔夫斯腾达姆大街和乔希姆斯达勒大街方向的效果图

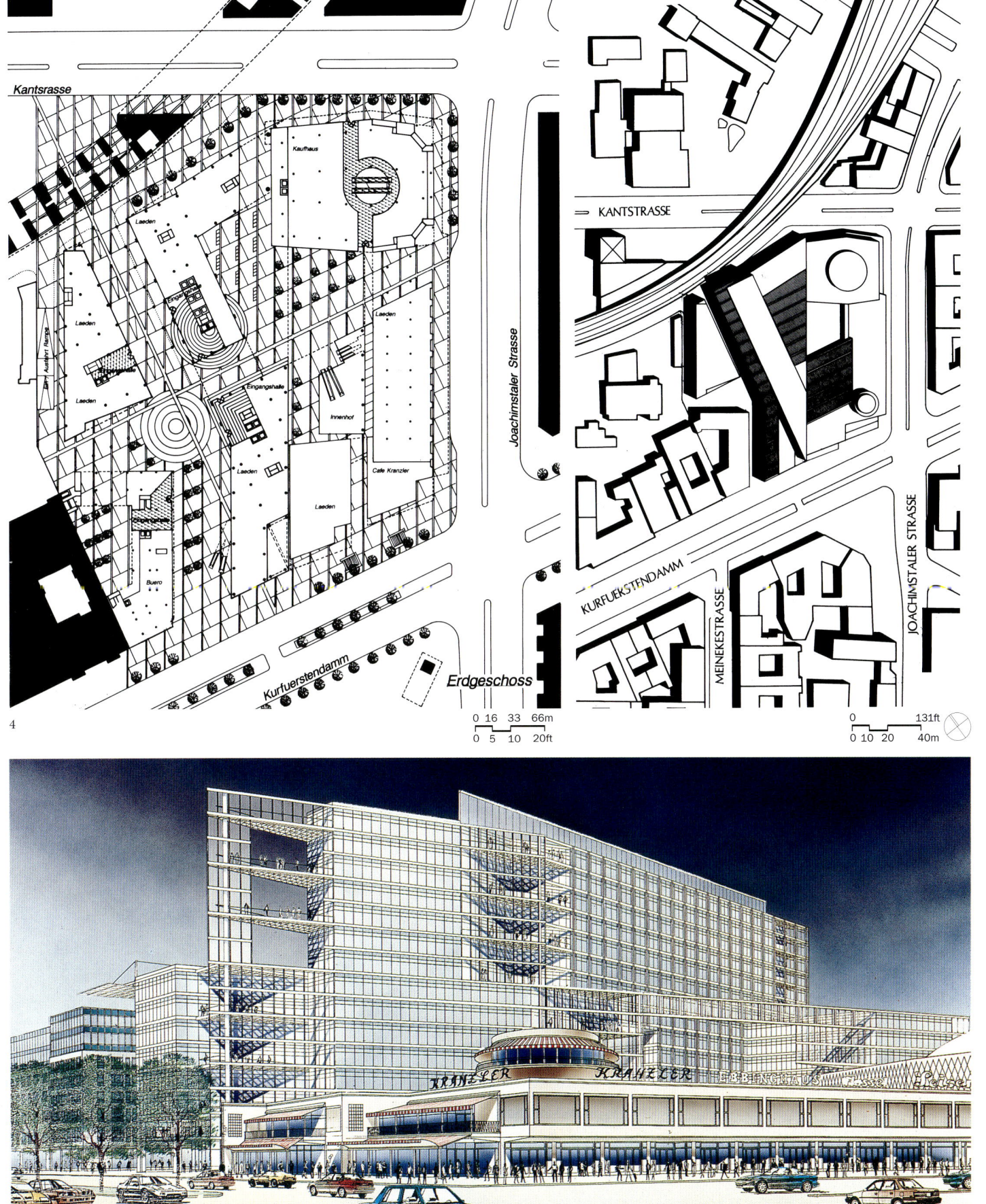

Ku-Damm 119

Design/Completion 1988/1995
Berlin, Germany
Athena Grundstuecks AG; Vebau GmbH Frankfurt/Berlin
150,000 square feet
Poured-in-place reinforced concrete
Panelized aluminum/glass curtain wall with green or gray tinted glass, patterned spandrel glass and painted aluminum panels; piers and base black granite

库-达姆119号

设计／竣工　　1988/1995
德国，柏林市
雅典娜地产股份有限公司，维堡股份有限公司（法兰克福与柏林）
150,000平方英尺
现浇钢筋混凝土
采用绿色或灰色玻璃的镶板式铝材幕墙和玻璃幕墙
压花层间玻璃及涂层铝材镶板；黑色花岗石柱子与基座

This building completes a partially existing block configuration. It adjusts to the irregular site with rounded corners acting like hinges, and steps down from eight floors to five floors. At the corners the building has special features, such as projecting glass bays, steel brows and steel masts, which reflect features of the historic city in a new way.

The linear nature of the building is reinforced through a banded aluminum and glass facade, which is replaced by glass at the two upper floors to correspond to the lower cornice height of the adjacent buildings. The building has a center core configuration allowing for flexible leasing arrangements and multiple-tenant use. Below grade are 180 parking spaces, and service and technical areas.

A landscaped courtyard provides accessible urban open space, offering additional urban linkages within this city quarter.

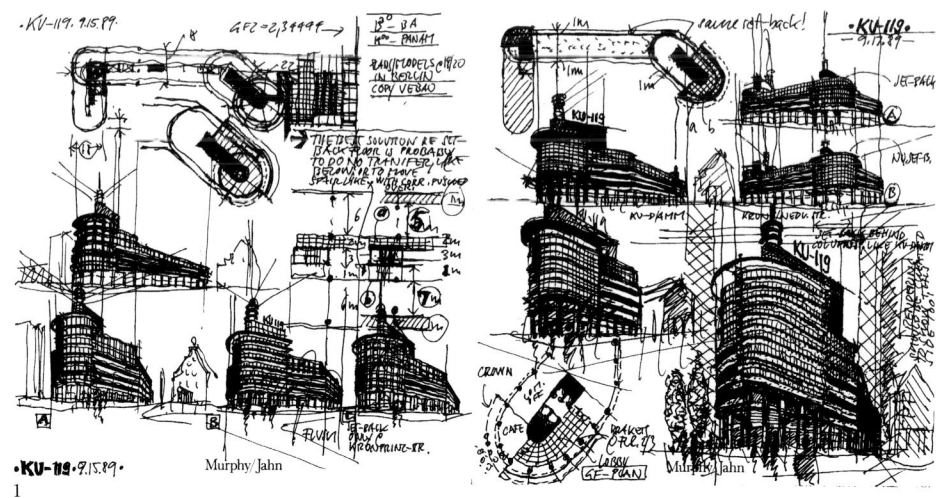

该项目完善了部分存在的街区构图形式。该建筑采用圆形转角处理，起着类似于铰链的作用，从而使其适应了不规则的建筑基地，并且从第八层以台阶的形式过渡到第五层上。在转角的地方，该建筑体现出某些独特的特色，例如，采用了凸出的大玻璃开间形式、钢制的檐板以及钢制立柱桅杆，所有这些细部处理都用一种新的方式反映出了这座历史名城的特色。

大楼的线型特征通过水平条带状的铝材和玻璃构成的立面处理得以强化，并且在顶部的两层上采用玻璃墙面，使该建筑与邻近建筑物的檐口高度相一致。该大楼采用了中央核心筒结构形式，可以实现灵活的租赁安排和满足多租户共同使用的需求。在地坪以下是可停放180辆车的停车场，还有设备和技术用房。一个绿化庭院形成了可以进入的城市开放空间，从而在这个城市区域中提供了更多的与城市连接的方式。

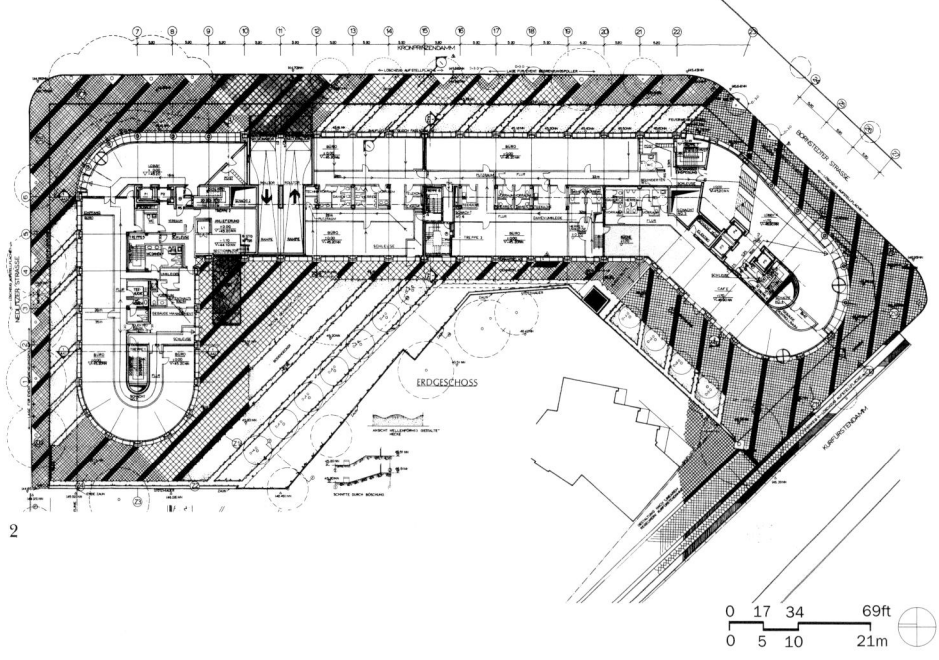

1　Concept sketches
2　Floor plan
3　Round corner/entrance
4　Banded curtain wall
5　Linear building expressing rounded corners/entrance

1　构思草图
2　楼层平面图
3　圆形转角处及入口
4　条带状幕墙处理
5　用线型建筑表达圆形转角处和入口

3

4

5

Ku-Damm 119 153

Mannheimer Lebensversicherung Headquarters

Design/Completion 1988/1992
Mannheim, Germany
Mannheimer Lebensversicherung AG
236,000 square feet
Concrete structure, steel brow
Curtain wall of glass, blue pearl stone and aluminum

曼海姆人寿保险公司总部大楼

设计/竣工　　　1988/1992
德国，曼海姆市
曼海姆人寿保险股份公司
236,000 平方英尺
混凝土结构，钢檐板
玻璃、蓝色珍珠岩及铝材幕墙

The brief required 150,000 square feet of office space (open-plan or single offices) to be used by the client and leased to tenants, as well as a cafeteria and parking for 280 cars.

Driving from the autobahn into Mannheim, the site is at the entrance to the inner city, at the beginning of the Augusta-Anlage—one of the most beautiful avenues in Germany. The design gives maximum presence to the building's corner site, guiding movement down the boulevard with an elegant curving gesture. The building form has a maximum height of 14 stories, accentuated by a cantilevered steel top, at the corner of the Augusta-Anlage. Moving along the boulevard towards the inner city, the building steps down into a series of two-story outdoor terraces to a typical height of four stories.

The ground floor contains the entrance lobby, dining rooms and an employee cafeteria that extends out into the landscaped courtyard in a gently curved form.

　　该项目的任务书要求设计 150,000 平方英尺的办公空间（包括开敞式或独立办公室式），供甲方自己使用或出租给客户，还包括一个自助餐厅和可停放 280 辆车的停车场。

　　开车沿高速公路进入曼海姆市，即可看见这个位于进入市区的入口处的建筑地块，这也是奥古斯塔市政大街的起点，这条大街是德国最优美的大街之一。该设计在最大限度上表现了建筑转角处的场地，用一个优雅的曲线形式来将人流引导到林阴大道上。该建筑在造型上最高处为 14 层，在奥古斯塔市政大街的转角处用一个悬挑出来的钢制顶部构架来增强最高部分的视觉效果。顺着林阴大道往市区方向，该建筑逐渐向下退台，形成了一系列的由两层楼高的室外平台，一直跌落至 4 层楼的高度上。

　　底层部分包括一个入口大厅、餐厅以及一个雇员自助餐厅，该餐厅通过一个柔和的曲线形式来延伸至中央的绿化庭院处。

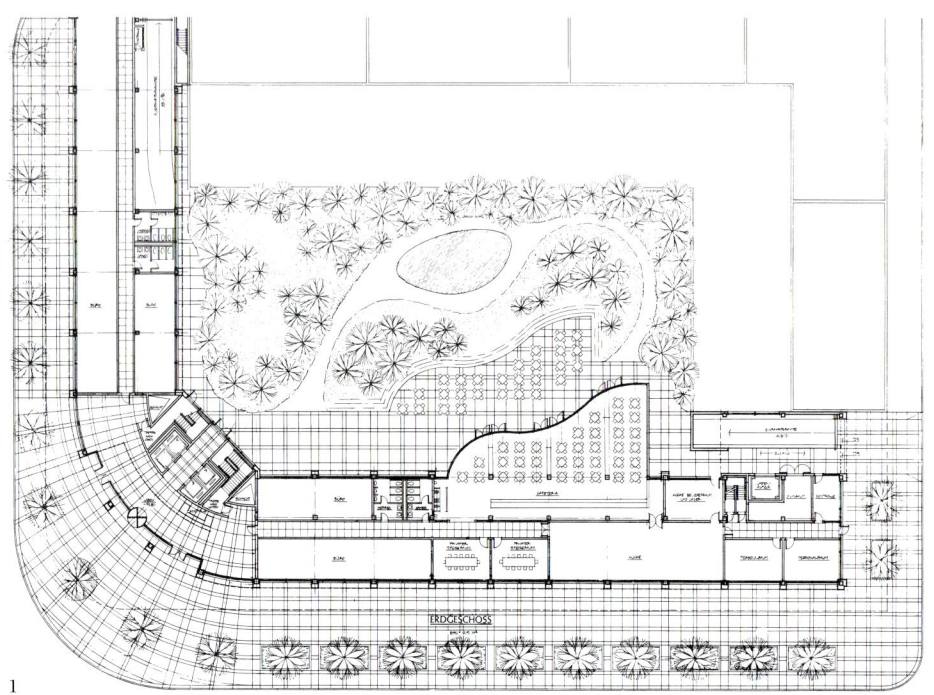

1

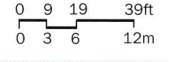

2

1　Site/ground-floor plan
2　North elevation
3　Landscaped courtyard
4　Building presence entering Mannheim
5　Cantilevered brow
6　Entrance

1　总平面图及底层平面图
2　北向立面
3　绿化庭院
4　在进入曼海姆市的道路上看建筑外观
5　悬挑式檐板
6　入口

3

4

5

6

7　Office corridor
8　Private office
9　Lobby
10　Cafeteria
11　Kneipe

7　办公室走廊
8　私人办公室
9　自助餐厅
10　小酒店

7

8

Mannheimer Lebensversicherung Headquarters

Augustinerhof-Nuernberg

Design/Completion 1990/1997
Nuernberg, Germany
DHF Bautraegergesellschaft
165,000 square feet
Concrete structure
Stone facade with punched window openings; glass and steel bays with painted aluminum mullion system; suspended and button-glazed galleria roof and retail bays

奥古斯丁霍夫-纽伦堡

设计/竣工　1990/1997
德国，纽伦堡市
DHF建筑公司
165,000平方英尺
混凝土结构
开点式窗洞口的石制立面，采用涂层铝材竖框系统的玻璃与钢制凸窗；悬挂式和点式玻璃节点的廊道屋面及商业零售门面

The goal of the design is to create a public covered passage between Winklerstrasse and Karlstrasse as a connection to the "Hauptmarkt". The different uses are stacked in a logical way: parking for 240 cars in two below-grade parking levels; food court and loading in the first below-grade level; retail and access to office, hotel and residential functions on the ground floor; and retail on the mezzanine level. On the first through fifth floors are located the 152-room hotel, offices, and residential use along the Augustinerstrasse.

该项目的设计目的是要形成一个在温克勒大街及卡尔大街之间的有顶通道，以之形成与"中心市场"的连接。不同的用途按照一种有逻辑的方式布置起来：地下有两层用于停车，共有240个停车位；地下第一层为餐厅和装卸区；地面底层设置了商业零售区和通往办公室、酒店及住宅区的通道；沿奥古斯丁大街方向上在第二层到第六层中设置了有152个房间的旅馆、办公楼和住宅功能部分。

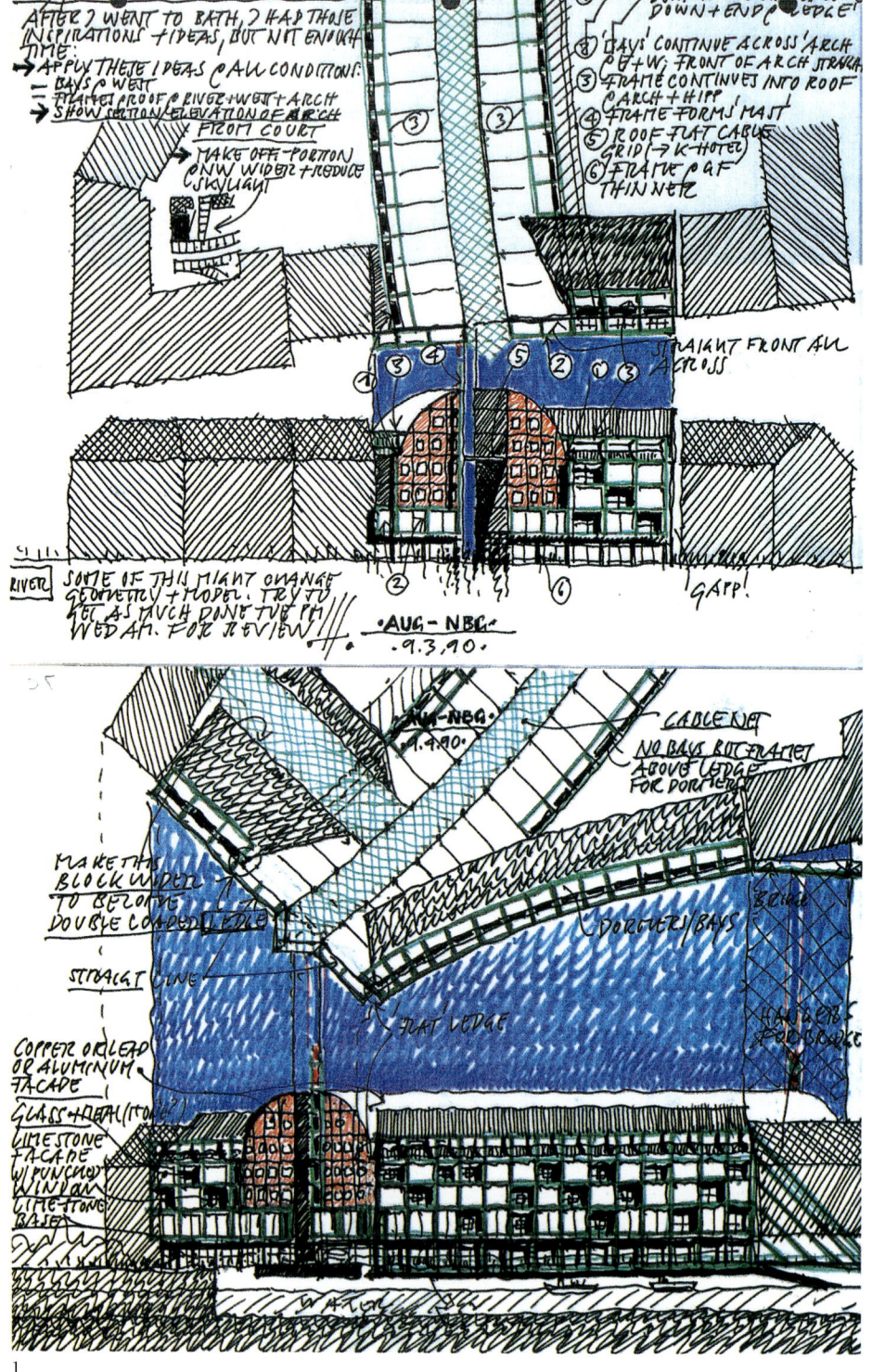

1

1	Concept sketches	1	构思草图
2	Entrance	2	入口
3	Site/floor plan	3	总平面图/楼层平面图
4	Pedestrian passageway	4	行人通道

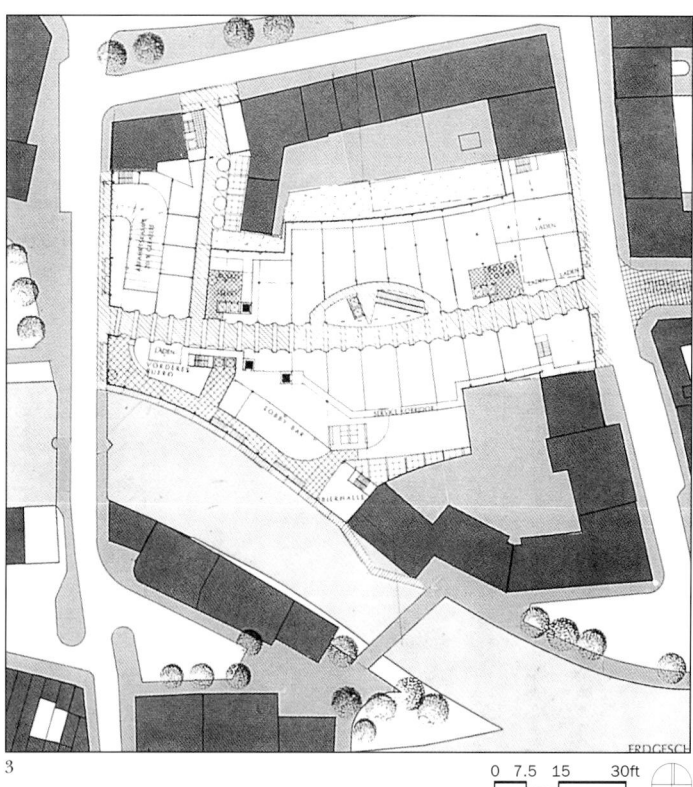

Augustinerhof-Nuernberg 159

Europa-Haus

Design 1992
Frankfurt, Germany
SkanInvest
1,094,170 square feet
Placed concrete; steel and glass galleria roof
Coated aluminum supergrid, tinted low-E glass and patterned fritted glass in painted aluminum mullion system; textured aluminum panel system

欧罗巴住宅

设计　　1992
德国，法兰克福市
斯坎投资公司
1,094,170 平方英尺
浇注式混凝土，钢材与玻璃廊道屋面
涂层铝材巨型栅格，彩色低弹性模数玻璃与压花热熔玻璃嵌于上漆铝材竖框系统中，带纹理的铝板系统

The design for Europa-Haus works within the typology of the low block-type structures of the European city. The building becomes an urban event and creates public open spaces and linkages in a city quarter which is presently devoid of those. Here, however, the typical block-type is reconfigured. The interior court is opened, "publicized", and transformed into a pleasant, eventful urban passage.

The building configuration is the antithesis of the typical office tower in that the buildings follow the streets and conform to the urban character of the area. The building creates at once a "gate" towards the main street and a spacial closure to the urban open space to the east.

The differentiated building massing and its low height create a transition of scale and a compatibility with the existing surrounding structures. This is aided by the treatment of the building massing with different forms and surfaces.

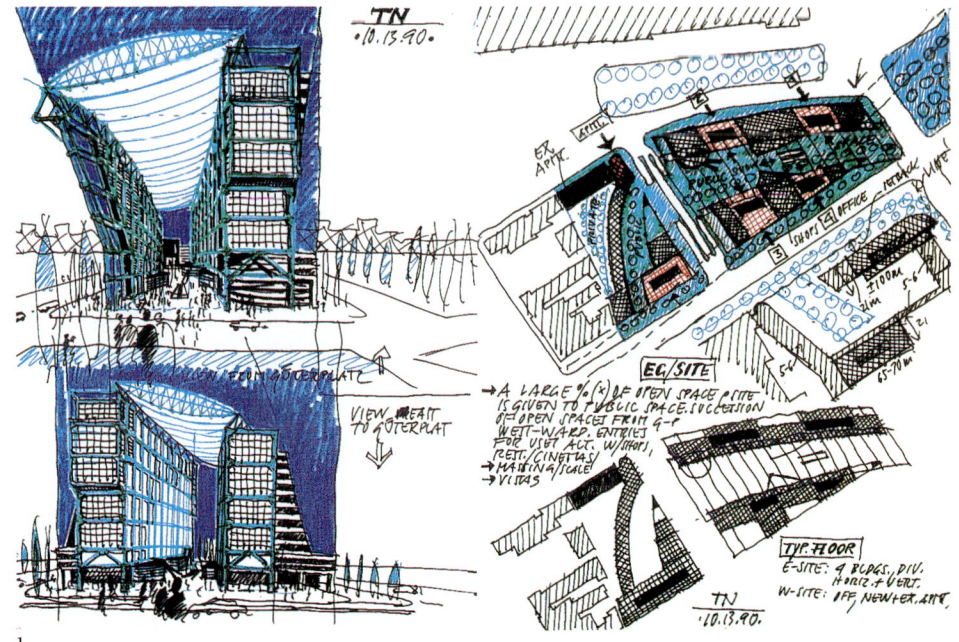

欧罗巴住宅的设计是在欧洲城市的低层街区建筑的类型上进行的工作。该建筑本身成为一项城市活动，也形成了在一个城市居住区中的一些公众开放空间和联系，通常居住区都缺少这样的一些空间和联系。但是，在这栋建筑上，典型的街区建筑形式被予以重新建构。内部的庭院是开敞式的，具有"公共性"的特征，并且被变形为一个令人愉悦的、场景式的都市通道。

该建筑在构成上与典型的办公楼的外观相反。在这个设计中，建筑物均沿街道布置，并与该地区的城市特色相一致。该建筑因此而形成了一个面朝主街道的大门和一个对位于东向的城市开放空间的空间围合。

具有不同特征的建筑体量处理及其低矮的建筑高度形成了对尺度的转变，也与现有周边建筑特色相协调。此外，还通过在建筑体量处理方面使用不同的形式和表面处理来使这一特色得以强化。

1　Concept sketches
2　Urban passage "gate", creating an interior open court

1　构思草图
2　城市通道"大门"，形成了一个内部的开放庭院

Lufthansa Corporate Headquarters

Design 1990
Cologne, Germany
Lufthansa German Airlines
539,636 square feet
Placed concrete, steel framed roof and masts
Patterned fritted vision glass and metal panels

汉莎航空公司总部大楼

设计　　1990
德国，科隆市
汉莎德国航空公司
现浇混凝土，钢框架结构屋面及桅杆
压花热熔观景玻璃窗及金属面板

The sketches of the Lufthansa headquarters are based on a "unit-type" building with central courtyard. This configuration was chosen above other building types as it best met the design criteria.

In plan, the building is basically a number of individual buildings, connected through courtyards and bridges. This allows the most compact arrangement of offices and work spaces and the most comfortable working conditions. Individual offices and working areas can be arranged either in the middle or on the edge of the modules. The inside areas which are flooded with daylight and enhanced by planting, create a separation from the bordering areas, but also make visual communication possible.

The image of the building is characterized by the wing-shaped roof, making the association with aircraft. The aluminum cladding and hi-tech detailing heighten the expression of technology and progress.

　　汉莎航空公司总部的草图设计是建立在一个带有中央庭院的"单元式"建筑的构思上的。选择这种构图形式而不是其他类型的原因是它最能够达到规定的设计标准。

　　从平面上来看，该建筑实际上是由一系列独立的建筑物而构成，通过庭院及桥梁将各个单体连接在一起。这种设计可以实现最紧凑的办公及工作空间布置和提供最舒适的工作条件。独立的办公和工作区域既可以安排在各个建筑组块的中间，也可以布置在其边缘部分。建筑群内部的区域具有充沛的日照，并用植被来强化其特征，从而形成了一种与边缘区域的分隔，但仍然可以实现视线上的联系。

　　该建筑在造型上以机翼状的屋面为主要特征，形成了与飞机相关的联想。铝材覆面处理和高科技的细部处理加强了在该建筑上对技术和时代进步的表达。

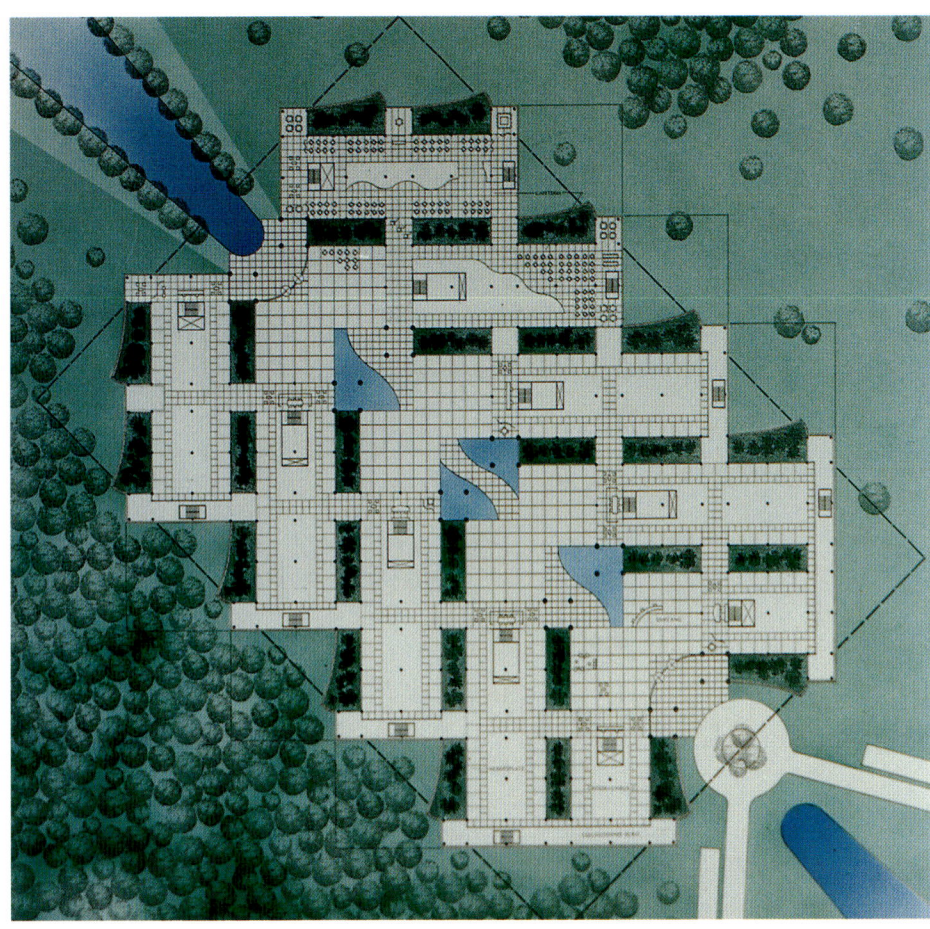

1

2

1　Ground-floor plan
2　Overall view

1　底层平面图
2　建筑全景

Harold Washington Library Center Competition

Design 1988
Chicago, Illinois
City of Chicago Library System; Tishman Midwest
720,000 square feet
Placed concrete, steel arches at vaulted roof
Black granite, gray granite, structural glazing, gray marble, terne metal roof

哈罗德·华盛顿图书馆中心设计竞赛

设计　　1988
伊利诺伊州，芝加哥市
芝加哥市图书馆系统；提什曼中西部机构
720,000平方英尺
现浇混凝土，拱形屋顶部分采用钢制拱架
黑色花岗石、灰色花岗石、结构部分玻璃墙面、灰色大理石、镀铅锡铁板屋顶

The site for the Harold Washington Library Center invites a strong urban planning solution. Recognizing this, the design overcomes the barrier formed by the Loop elevated tracks. Rather than isolating the library to either north or south of Van Buren, the main part of the building is raised above the tracks.

The most public functions of the library—the museum store, the restaurant, and the bookstore—are located at street levels. The auditorium and exhibit halls are located on a more formal streetfront mezzanine, thus preserving the retail and entertainment character of State Street.

The non-secure public functions are housed on four large floors spanning Van Buren, linked to the rest of the library through a large central open space. This space acts as an internal street, organizing and unifying the various elements while providing for a soft, natural and indirect light that diffuses throughout the building.

哈罗德·华盛顿图书馆中心的建筑基地需要用一个极富技巧的城市规划方案来处理。在认识到这一点后，该设计克服了因为环城高架轨道的存在而产生的障碍。为了使该图书馆不会被孤立地单独设置在凡·布伦轻轨线的北侧或南侧，该建筑的主体部分被抬高架设在轨道的上面。

图书馆最重要的公众服务功能区（包括博物馆展厅、餐馆及书店）均位于与街面相平的首层。报告厅和展览大厅则设置在更加整齐匀称的临街夹层里，从而保护了州立大街上所具有的商业零售和娱乐建筑的特色。

一些不需要保安的公共性功能被安排在横跨瓦·布伦轻轨线上的四层巨大的楼层里，通过一个大型的中央开放空间将它与图书馆的其他部分联系起来。该空间起着内部街道的作用，将各种不同的建筑元素组织和统一起来，同时也提供良好的采光，在整个建筑里都弥漫着柔和的、自然的和间接式的折射光线。

1　构思草图
2　州立大街方向的立面
3　从东南方向看建筑

1　Concept sketches
2　State Street elevation
3　View from the south-east

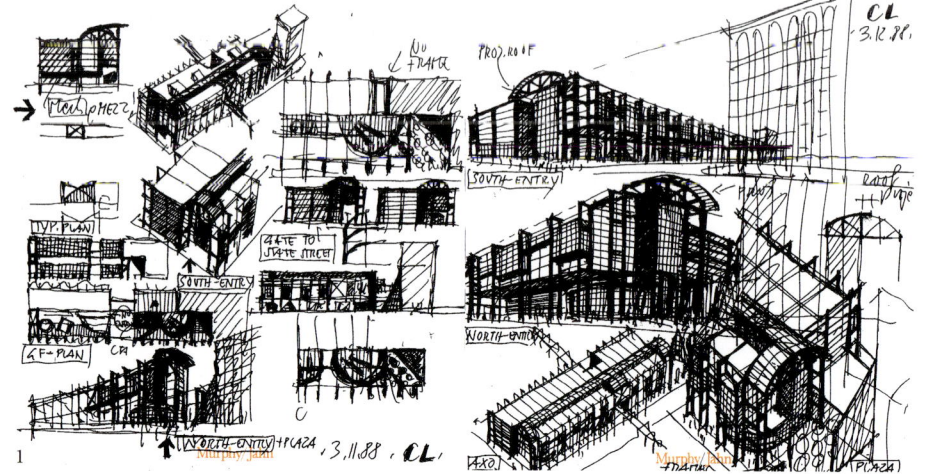

First & Broadway Los Angeles Civic Center

Design 1987
Los Angeles, California
RCI
630,000 square feet
Steel frame
Stone-clad columns, infill of aluminum grid sun shade and fritted glass

弗斯特与百老汇洛杉矶市政中心

设计　　1987
加利福尼亚州，洛杉矶市
RCI 公司
630,000 平方英尺
钢结构
柱子石材贴面，铝材栅格式遮阳板和热熔玻璃填充墙

The First & Broadway development is a much-needed component in the completion of the overall concept of the Los Angeles Civic Mall.

The office building can be interpreted as "Siamese twins"— two 18-story buildings joined at the top providing a combined total of 600,000 gross square feet of office space. This unique building concept creates a monumental portal to the central plaza and a covered area for building entrances, while providing more functional, leasable floor areas.

The open spaces associated with the site are designed to respond to local criteria as well as large-scale urban issues. The siting of the building mass creates the formal urban plaza to the north. A variety of sizes and uses of open space are offered to the public: the grand central civic space; the more informal, smaller-scaled open spaces adjacent to the ground floor of the building; galleria and arcades associated with the retail; and rooftop terraces for daycare and dining.

在总体设计规划竞标中，弗斯特与百老汇开发项目是在实现洛杉矶市政中心的总体构思中非常重要的一个组成部分。

该办公大楼可以被看成"连体双胞胎"——两栋18层高的大楼在顶部被联系在一起，从而形成了一个总共达600,000平方英尺的办公空间。这一独特的建筑构思为中央广场形成了一个纪念碑式的桥形入口，并覆盖着大楼的数个大门部分，同时也提供了更多的可使用和出租的建筑面积。

与该建筑基地联系在一起的开放空间在设计上既满足了当地的建筑标准，也与大尺度的城市特色形成呼应关系。建筑体量在场地上的布置方式在其北侧形成了一个规整的城市广场。各种尺度和用途的开放式空间向公众开放：大型中央市政空间；一些不太规整的、尺度较小的开放式空间与建筑的底层联系在一起；与商业零售区相连的商业街廊和临街廊道；以及用于日托和餐饮服务的屋面平台。

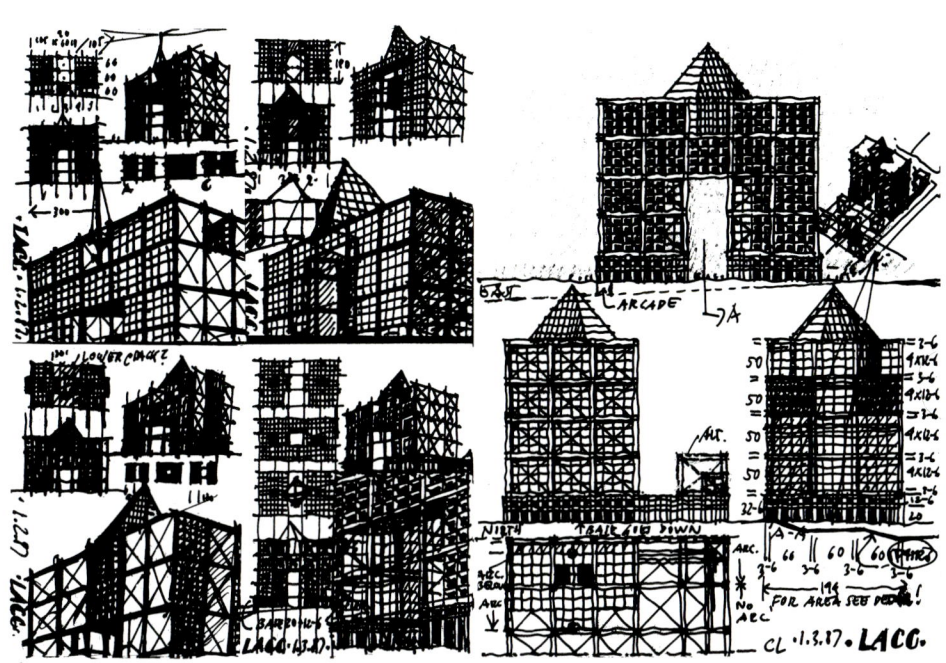

1　Concept sketches
2　Public plaza and building with portal, showing major and minor mullion systems

1　构思草图
2　拥有桥门入口的公共广场和建筑物，可以看到主要的和次要的竖框体系

State of Illinois Center

Design/Completion 1979/1985
Chicago, Illinois
State of Illinois; Capital Development Board
Joint Venture Architects: Lester B. Knight & Associates
1,150,000 square feet
24,000 square foot skylight over 160 foot diameter atrium
Structural steel frame, composite deck
Aluminum mullions with clear and reflective glass

伊利诺伊州立中心

设计/竣工　　1979/1985
芝加哥，伊利诺伊州
伊利诺伊州，首府开发委员会
合作建筑设计单位：莱斯特·B·耐特及联合事务所
1,150,000平方英尺
2,400平方英尺的采光天窗架设在直径160英尺的中庭上空
钢结构，复合楼板
铝材竖框，采用透明和反射玻璃

The new State of Illinois Center responds to and reinforces existing urban patterns. Its two main components are a low block structure and a large central rotunda.

The south-east portion of the block is sliced away to form a sloped setback configuration with open space at the corner. The continuous but stepped-back west facade of the building rises from the street line to reinforce the LaSalle Street Canyon. An indentation of the skin at the lowest two floors creates a covered arcade which continues along LaSalle and Clark Streets as well as along the curved south-east wall.

The new element in the State of Illinois Center is the reading of the central space from the outside. This element of openness is continued along the curved facade by the five-story atria which follow the setbacks.

The interior rotunda space is a logical alternative to the exterior open spaces which already characterize other city plazas along the Dearborn-Clark corridor.

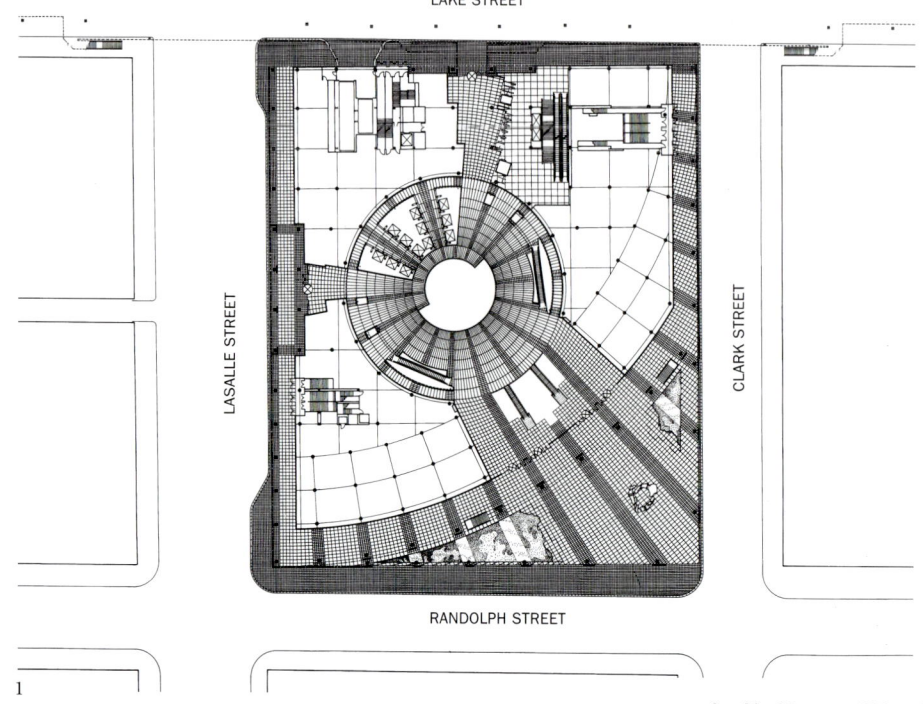

新规划的伊利诺伊州立中心与现有的城市模式相呼应并进一步予以强化。该项目有两个主要组成部分，一个是低层的街区建筑，另一个是巨大的中央圆形建筑物。

街区建筑的东南部分被切掉，形成了一个倾斜退台式的构图，在转角处形成了开放式空间。建筑物连续的但向后退台式的西立面从街道标高处升起来，强化了拉萨勒大街上的街道空间。在最下面两层的缺口形状的外墙部分形成了一个有顶盖的廊道，这种处理在拉萨勒大街和克拉克大街上连续延伸，也出现在曲线型的东南向外墙上。

在伊利诺伊州立中心上采用的新手法是可以从建筑物外部来解读其中央空间。这一开放性的元素通过顺着由退台处理而形成了5层高的中庭所形成的曲面式的立面而得以持续体现。

内部的圆形大厅空间是其外部开放空间的一种有着逻辑关系的结果，这种外部空间的形式成为沿着迪尔伯恩－克拉克走廊上的其他城市广场的主要特征。

1　Site/ground-floor plan
2　City context/east elevation
3　Atrium

1　总平面图/底层平面图
2　城市文脉/东向立面
3　中庭空间

4

5

6

4	View from Clark/Randolph Streets	4	从克拉克/伦道夫大街看建筑
5	Pedestrian arcade/plaza	5	步行廊道
6	Interior atrium	6	室内中庭
7	Ground-floor atrium with retail	7	设置有商业零售区的底层中庭
8	Entry gate	8	入口大门

Shand Morahan Corporate Headquarters

Design/Completion 1982/1984
Evanston, Illinois
Walken Souyoul Interests
173,000 square feet
Steel frame
Green fluoropolymer-coated aluminum curtain wall system for reflective glass, gray and black granite cladding

桑德·莫拉罕集团公司总部大楼

设计/竣工　1982/1984
伊利诺伊州，埃文斯顿
沃尔肯·苏尤尔股份公司
173,000平方英尺
钢结构
绿色氟聚合物－涂层铝材幕墙系统，反射玻璃，灰色和黑色花岗石贴面

This is the first phase of a two-phase project for the headquarters of an international insurance underwriter. Phase I has 173,000 square feet in eight stories, Phase II has 150,000 square feet in six stories.

The building is located on the edge of the central business district. An elevated railroad embankment borders the site and creates its triangular shape.

The L-shaped building was designed to provide a building not wider than 90 feet with a maximum distance of 45 feet to a window from any point within the building. The L shape also provides an opportunity to "reach out" and provide a covered arcade for the parking side of the building as well as the side toward the commuter facilities.

The entrance is located at the corner to provide optimum access for employees using either parking or commuter facilities. Executive offices are located on the top floors and access roof terraces.

　　该建筑是一个分为两期的项目中的第一期工程，该项目是一家国际保险公司的总部大楼。第一期工程为8层楼，建筑面积为173,000平方英尺，第二期为6层楼，建筑面积为150,000平方英尺。

　　该建筑位于中央商务区的边缘处。一条高架铁路的路基与该建筑基地接壤，并形成了三角形的基地。

　　该建筑被设计成"L"型，形成了一个宽度不超过90英尺的建筑，并且从建筑内部的任何一个地方到窗户的最大距离不超过45英尺。"L"型的平面设计也形成了一种"向外伸出"的机会，并在建筑的停车场一侧和面向通勤车站设施的一侧形成了一个连续的有顶盖的廊道。

　　入口大门位于建筑的转角处，为要使用停车场或通勤设施的工作人员都提供了最佳的行进路线。行政办公室位于最上面的几个楼层，并可通往屋面的平台。

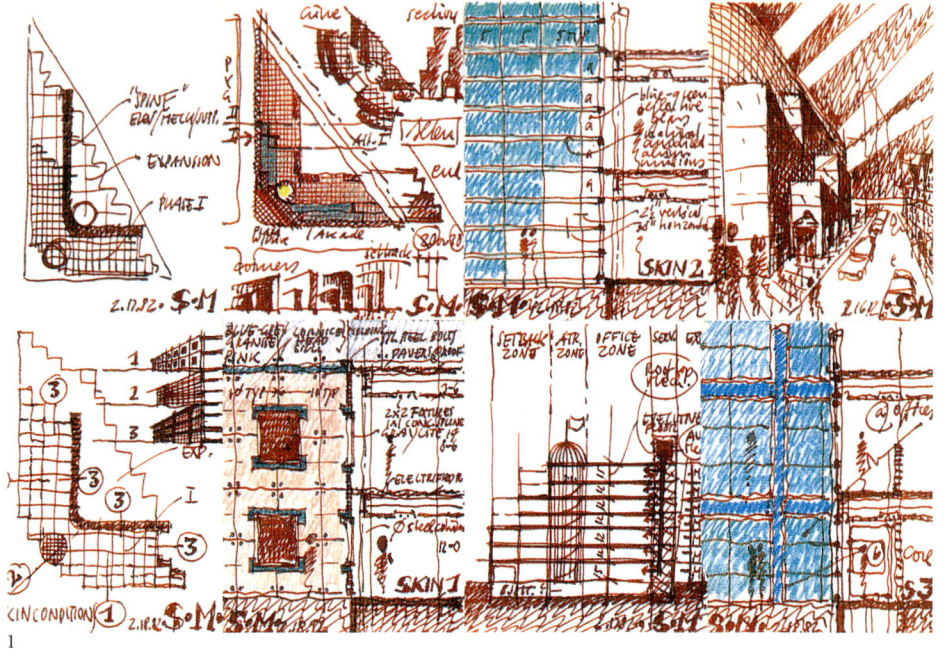

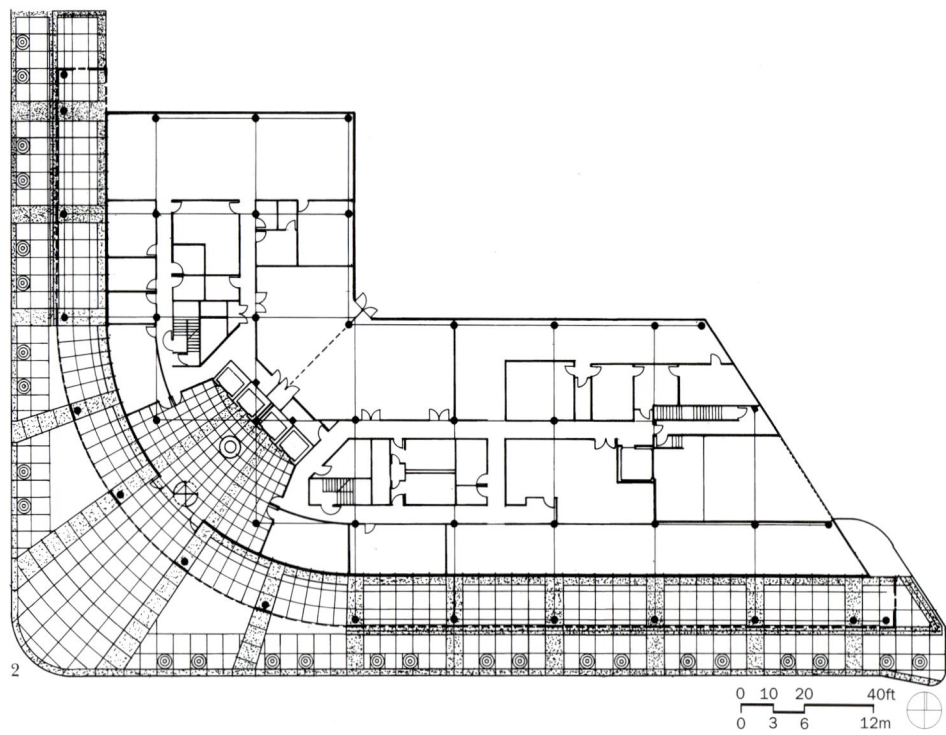

1　Concept sketches
2　Site/floor plan
3　Corner orientation with inner and outer skins
4　Chamfered end wall
5　Pedestrian arcade with punched windows in granite wall

1　构思草图
2　建筑基地/平面图
3　建筑转角处的内层和外层墙面处理
4　斜切的山墙部分
5　人行廊道与花岗石墙面上的窗洞

Shand Morahan Corporate Headquarters 169

First Source Center

Design/Completion 1978/1982
South Bend, Indiana
First Source Bank; Rahn Properties; City of South Bend
490,000 square feet
Cast-in-place concrete building frame; steel joint atrium roof frame
Clear and reflective insulated glass and metal panels

费斯特·索尔斯中心

设计/竣工　1978/1982
印地安那州，南本德市
费斯特·索尔斯银行；拉恩房地产公司；南本德市
490,000平方英尺
现浇混凝土建筑框架，钢节点中庭屋面框架
透明和反射绝缘玻璃与金属镶板

First Source Center is a multi-use urban development. It occupies a whole city block next to the new Century Center and acts as a catalyst for expansion of the CBD to the north. It consists of a 150,000 square foot bank/office building (First Source Bank), a 300-room hotel, a 27,000 square foot public atrium, and a 520-car parking garage.

The various functions have been distributed to reinforce existing compatible land uses. First Bank occupies the pivotal corner towards the CBD, while the hotel is oriented towards Century Center and contains provisions for a future closed pedestrian bridge. The atrium, which includes the hotel restaurant, is situated between those structures, and parking is on two levels below grade. From the enclosed atrium, a landscaped area with convenience parking provides the visual and physical continuity to a city park to the north of the site.

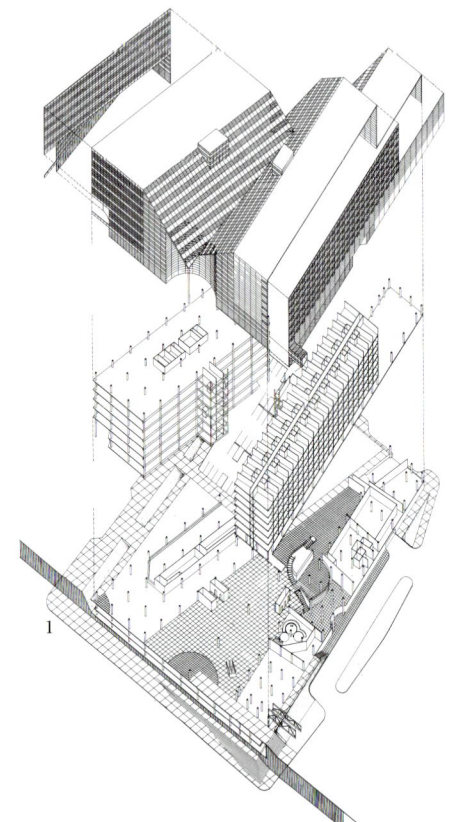

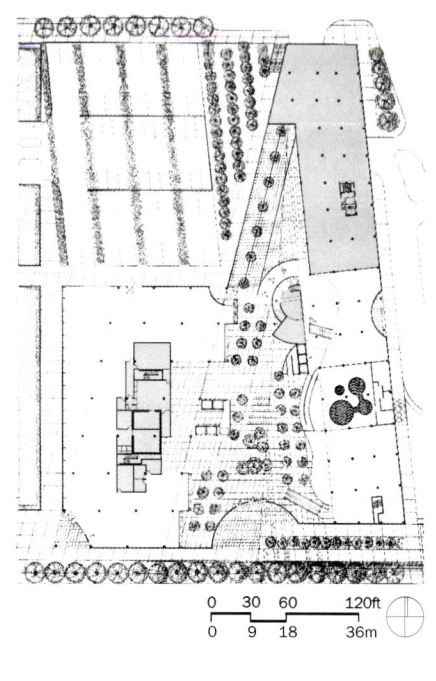

　　费斯特·索尔斯中心是一个多用途城市开发项目，它占据了位于新世纪中心旁边的一个完整的城市街区，并且成为中央商务区向北面扩展的催化剂。该项目包括一个面积为150,000平方英尺的银行和办公楼建筑（费斯特·索尔斯银行）、一个有300个房间的酒店、一个27,000平方英尺的公共中庭以及一个可停放520辆车的停车库。

　　建筑物的各种功能被分散布置，以强化已有的综合性的土地利用。费斯特·索尔斯银行位于朝向中央商务区的最重要的转角处，酒店则面对世纪中心，并且今后可以加建一座封闭式的人行天桥。设有酒店餐饮的中庭位于两个建筑体量之间，而停车库设置在地下的两层中。从封闭的中庭出去，有一个绿化区，设置有临时停车场，形成了与在该建筑基地北面的一个城市公园之间在视觉上和实际上的连续性。

1 Axonometric
2 Entrance detail
3 Front elevation
4 Atrium spine
5 Atrium entrance

1 轴测图
2 入口处的细部
3 正立面
4 中庭内景
5 中庭入口

3

4

5

First Source Center 171

作品精选

Selected and Current Works

交通建筑

174	凯宾斯基大酒店/慕尼黑机场中心大楼/人行通道
190	迪尔伯恩火车站
192	曼谷国际机场
196	科隆/波恩机场第二航站楼
200	戴高乐机场凯悦大酒店
204	阿卜杜勒阿齐兹国王国际机场设计竞赛
206	阿卜杜勒阿齐兹国王国际机场概念设计
208	第一美利坚广场/有轨电车车站
210	J·F·肯尼迪机场联合航站楼
214	联合航空公司第一航站楼
220	奥黑尔高速交通中转站

Transportation Buildings

174	Hotel Kempinski/Munich Airport Center/Walkways
190	Dearborn Station
192	Bangkok International Airport
196	Flughafen Köln/Bonn Terminal 2
200	Hyatt Regency Roissy
204	King Abdulaziz International Airport Competition
208	One America Plaza/Trolley Station
210	JFK Consolidated Terminal
214	United Airlines Terminal One Complex
220	O'Hare Rapid Transit Station

Hotel Kempinski/Munich Airport Center/Walkways

Design/Completion 1989/1994 (hotel), 1990/1996 (MAC), 1991/1992 (walkways)
Munich, Germany
Flughafen München Gesellschaft
Hotel: 350,000 square feet
MAC: 1,000,000 square feet
Hotel: Concrete structure, steel and glass roof supported by industry steel arches, cable-stayed glass end walls
MAC: Concrete structure, cable-stayed glass and steel roof
Walkways: Metal and glass roof supported by industry steel arches on steel columns
Hotel: Glass and aluminum curtain wall
MAC: Glass and aluminum curtain wall

The master plan for the Neutral Zone of the Munich Airport foresees the development of a 3,230,000 square foot service center and 9,000 parking spaces. In contrast to the flight-related buildings of the existing terminal, the buildings of the Neutral Zone will be dedicated to service uses, entertainment, relaxation, shopping and cultural events, as well as business endeavours.

The heart of the plan is the Munich Airport Center (MAC) which will establish a dominant east–west axis linking the existing Terminal West with a future Terminal East. A second axis orders all the uses in the Neutral Zone and runs through the center of the Hotel Kempinski. In its final phase the MAC will link the airport to the local and long-distance train networks.

The Hotel Kempinski is the first building in the Neutral Zone. The plan organization of the hotel corresponds to the airport level system: pedestrians circulate at level 03, and the roads and drop-off occur above at level 04. The level 03 entrance lobby can be reached either

Continued

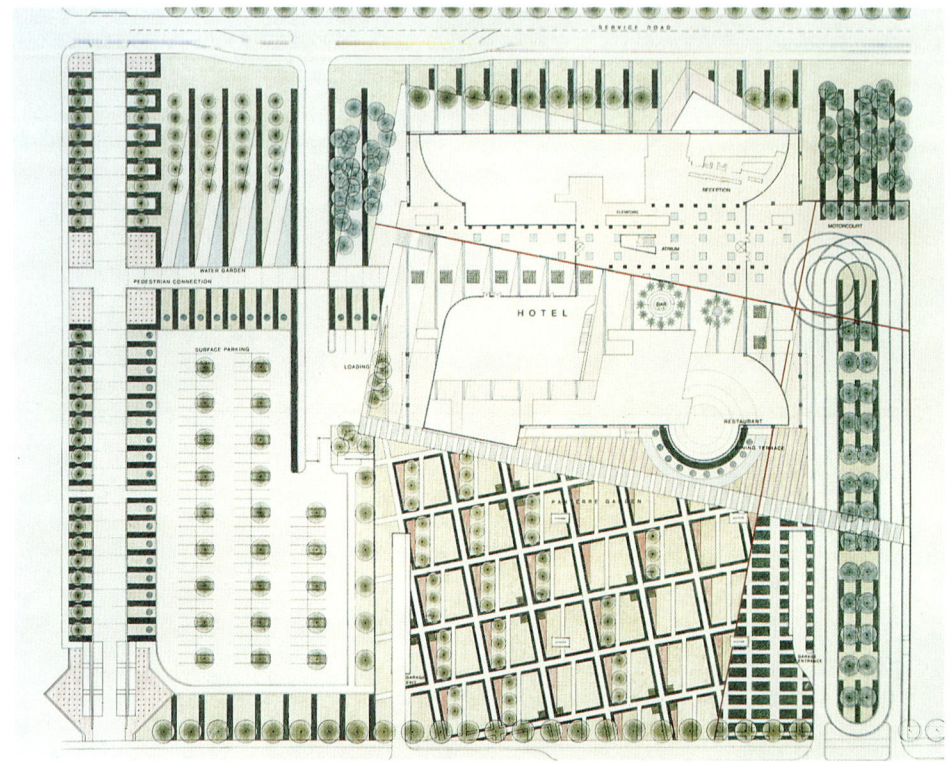

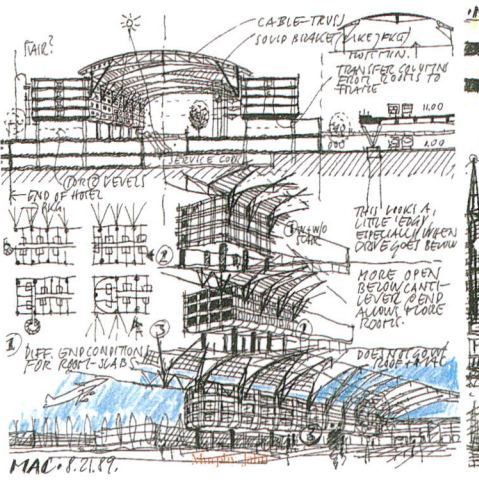
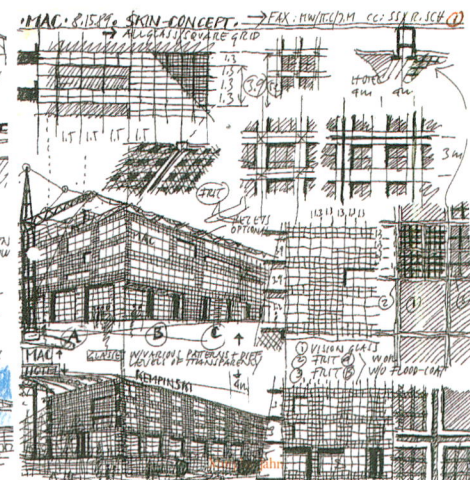

1 Site/landscaping plan
2 Concept sketches
3 Hotel Kempinski, covered walkways, and control tower
4 Aerial view
5 Landscaping

3

4

5

Hotel Kempinski/Munich Airport Center/Walkways 175

6

from the main terminal via the covered walkways, or from the below-grade parking structure in front of the hotel.

A glass stair and elevators lead to the main atrium at level 04. In concept, the atrium is both a private space for the hotel as well as a public reception hall for the expanding airport city. The transparency of the cable-supported glass end walls provides visual continuity between the drop-off, atrium and *biergarten* terrace extending to the south.

Covered moving walkways connect the existing terminal with the Hotel Kempinski, the MAC and bus terminal, and also the High Security Terminal along the eastern apron. In the east–west direction, the walkways extend the existing level 03 circulation routes of the terminal out into the open landscape, and in the north–south direction, they align with the longitudinal axis of the Neutral Zone centered on the atrium of the Kempinski Hotel.

Continued

7

楼通过带顶盖的人行通道到达，也可以从位于酒店前的地下停车库上来。

一个玻璃楼梯和几部电梯将人引向位于04水平面上的主中庭。在设计构思上，该中庭既可作为旅馆的专门空间，又可作为正在扩建中的机场城市的公共接待大厅。钢缆支撑的玻璃山墙所体现出来的透明性形成了在下车区、中庭和向南延伸的啤酒花园台地之间的视觉上的连续感。

有顶盖的人行通道将已有的航站楼与凯宾斯基大酒店、慕尼黑机场中心大楼和公共汽车终点站以及沿着东面裙房布置的安全检查航站楼等连接在一起，在东西方向上，人行通道将现有的位于03水平面上航站楼里的人流路线延伸出来，一直通到一片开放式的绿地上，而在南北方向上，建筑物与缓冲区的长向轴线平行，以凯宾斯基大酒店的中庭为该轴线上的中心。

（待续）

8

9

10

11

6	South elevation	6	南立面
7	Entrance/car approach	7	入口／车道
8	West facade/restaurant/garden walk	8	西立面／餐厅／花园小径
9–11	Curtain wall/roof details	9–11	幕墙／屋面细部
12	Elevation/section	12	立面／剖面
13	Hotel landscaping	13	酒店景观绿化
14	Sheltered entrance	14	带屋面的入口

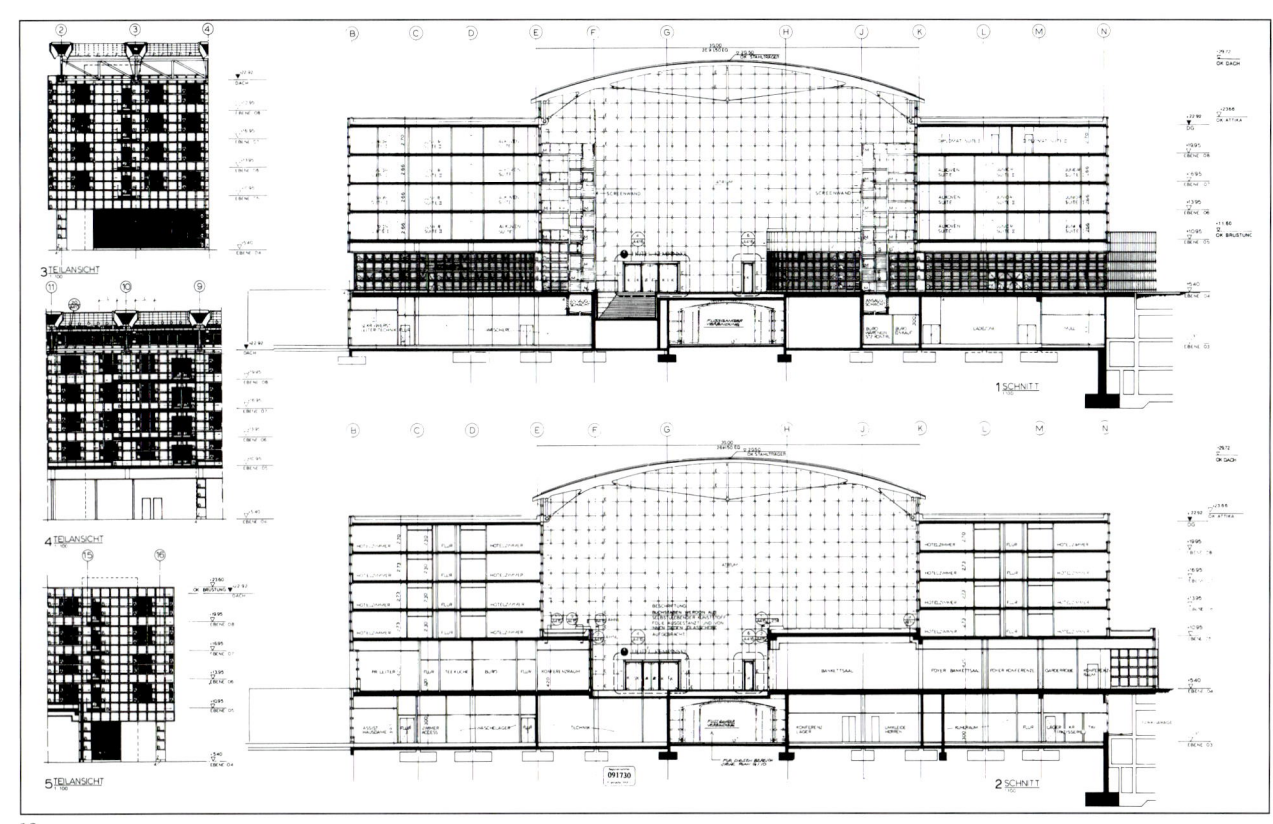
12

13

14

Hotel Kempinski/Munich Airport Center/Walkways 177

The walls of the walkways consist of curved, punched steel plate that is 50 per cent open and visually transparent, yet still solid enough to block strong winds and driven rain. The underslung steel roof structure mirrors the crossing arches of the Hotel Kempinski and MAC roofs.

Flanking the walkways are linear gardens. Hedges, interspersed with pyramidal evergreens are arranged perpendicular to the walkway structure, heightening the sense of motion. Where the walkways cross one another or enter the terminal, groves of tightly spaced birches set in bright orange crushed brick create points of transition and orientation. At these "decision points" the punched steel walls are replaced by full-height glass panels, letting in more daylight and affording clear views into the birch groves and the landscape beyond.

人行通道的墙面由曲面的和开了窗洞的钢板构成，其中有50％是开敞的，在视觉上是透明的，但仍然有足够的强度来抵挡狂风暴雨。向下悬垂的钢制屋顶结构延续了在凯宾斯基大酒店和慕尼黑机场中心屋面上的交叉拱的形象。

在人行通道的两侧是呈直线型的花园。点缀着一些锥形常绿植物的树篱与人行通道部分相垂直，从而加强通道的动感特征。在人行通道彼此交叉或进入到航站楼的地方，由被种植在明亮的橙黄色碎砖里的间距密集的桦树林起着转换点和指示运动方向的作用。在这些"具有决定意义的交点"处，开窗洞的钢墙面被通高的玻璃板所替代，可以使更多的光线进入到建筑之中，从而可以清楚地欣赏到桦树林以及在远处的绿地的美景。

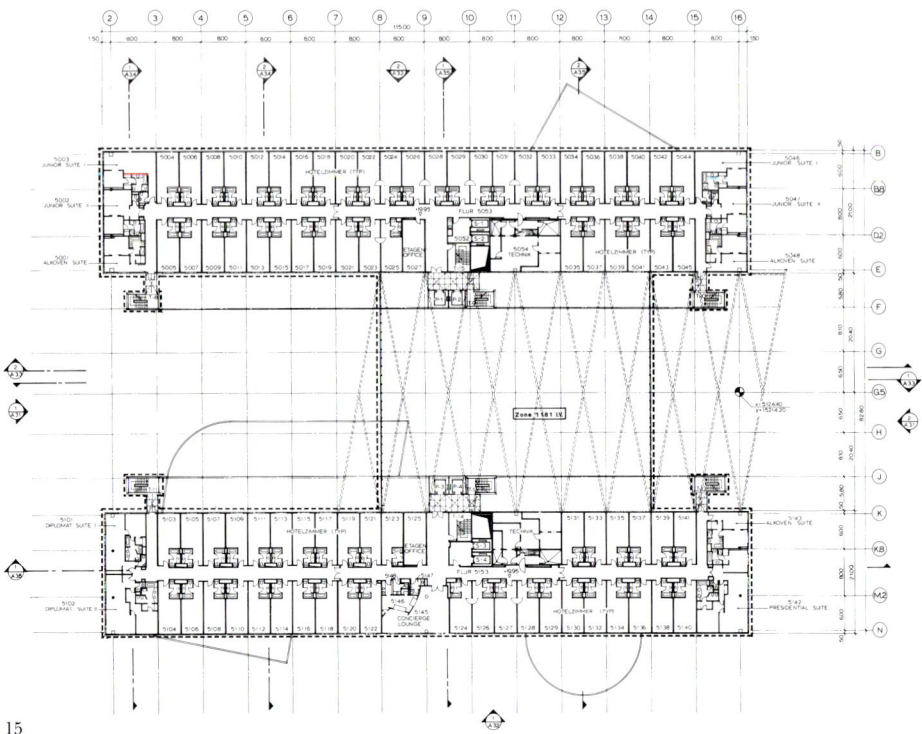

15

16

15	Typical hotel room plan	15	酒店房间标准层平面
16	Night view of atrium	16	中庭夜景效果
17–18	Stair/glass wall	17-18	楼梯/玻璃墙面
19	Geranium wall	19	天竺葵形成的墙面
20	Glass stair, lower level	20	玻璃楼梯及下部水平楼面

178

17

18

19

20

Hotel Kempinski/Munich Airport Center/Walkways 179

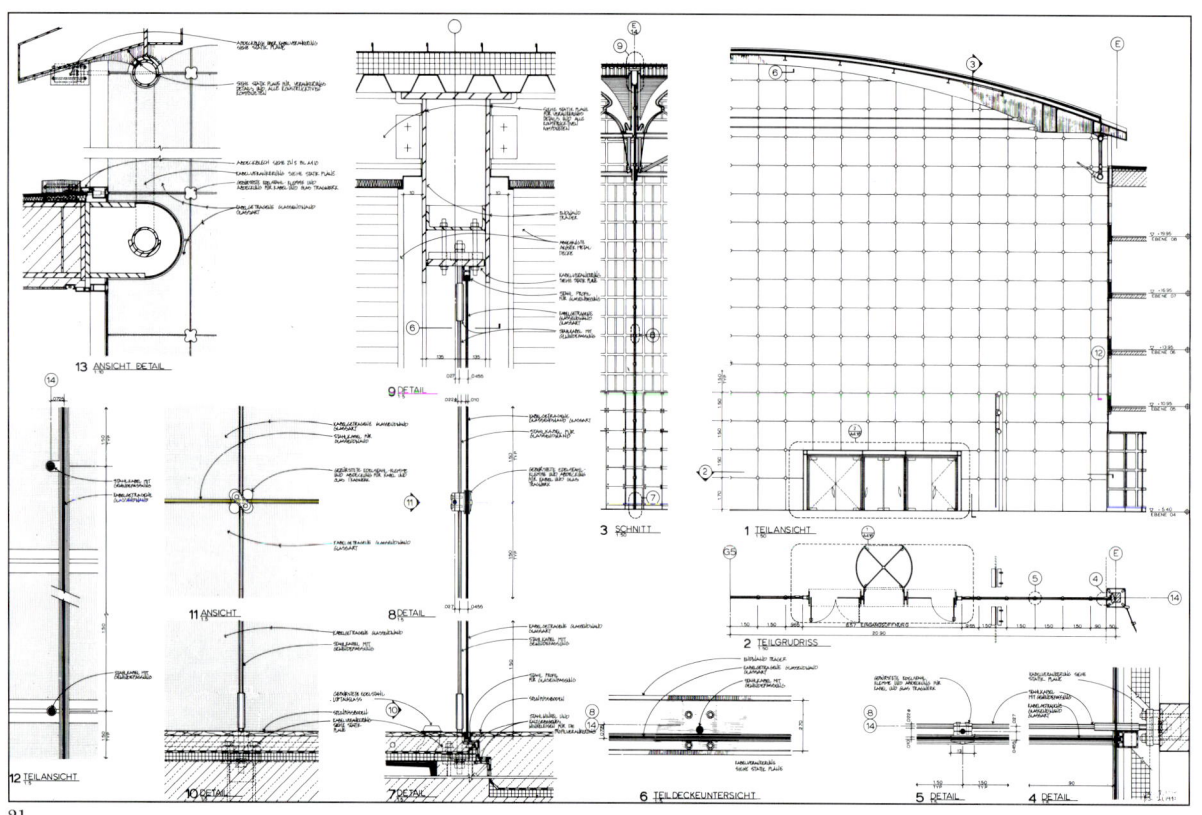

21 Elevation/plan/section: cable wall
22 Glass end wall
23 Terrace garden
24 Entrance wall
25 Flower wall

21 钢缆支撑墙：立面图/平面图/剖面图
22 采用玻璃的山墙面
23 退台形成的墙体
24 入口墙体
25 花之墙

23

24

25

Hotel Kempinski/Munich Airport Center/Walkways

26	Terrace	26	屋面平台
27	Glass flower wall	27	用玻璃做的花之墙
28	Stair, tree and flower wall	28	楼梯、树木和花之墙
29	Glass flower wall dividers	29	用玻璃做的花之墙分隔物
30	Stair cores	30	楼梯筒体
31	Glass pattern wall	31	压花玻璃形成了墙体
32	Tree detail	32	树木种植细部
33	Flower wall detail	33	花之墙的细部

26

27

28

29

Hotel Kempinski/Munich Airport Center/Walkways 183

34

35

36

37

38

34	Stair	34	楼梯
35	Cable wall connection	35	钢缆支撑墙的连接细部
36	Walkway viewed from above	36	从上面俯视步行通道
37	Fritted pattern glass	37	热熔压花玻璃
38	Stair	38	楼梯细部
39	Underground walkway to garage	39	通往停车库的地下步行通道
40	Walkway at night	40	步行道的夜景效果
41–42	Walkway connection to hotel	41–42	通往酒店的步行通道节点

39

40

41

42

Hotel Kempinski/Munich Airport Center/Walkways 185

43

44

43　Munich Airport Center
44　Interior MAC pavilion
45　Concept sketches: MAC
46　Glass stair from walkway level to atrium
47–49　Walkway details

43　慕尼黑机场中心
44　慕尼黑机场中心室内休息厅
45　构思草图
46　从人行通道层到中庭层的玻璃楼梯
47–49　步行通道细部处理

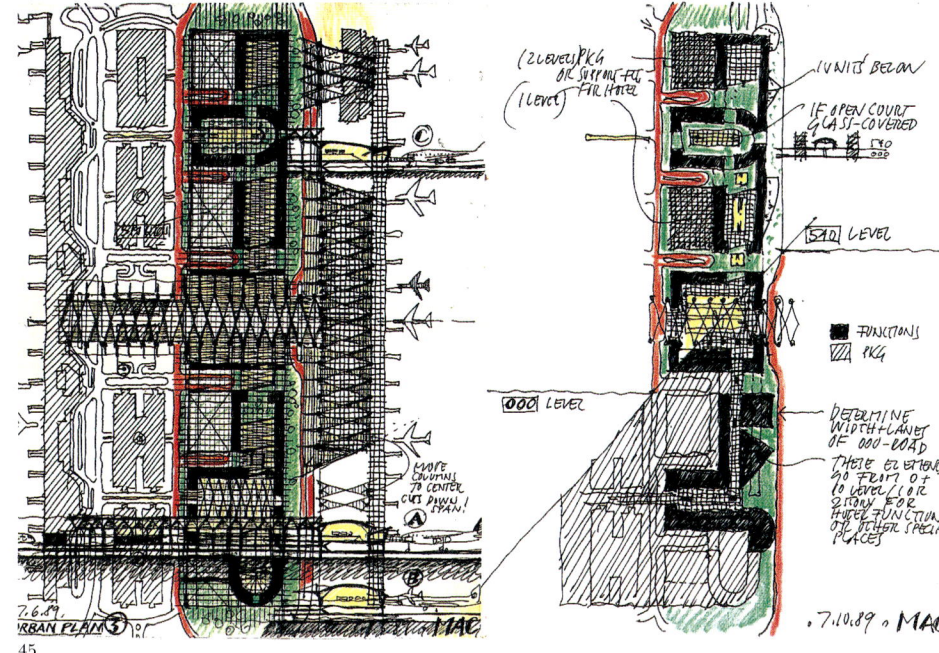

45

46

47

48

49

Hotel Kempinski/Munich Airport Center/Walkways　187

50	Walkway detail	50	步行通道大样图
51–52	Perforated metal detail	51–52	穿孔金属板细部
53	East–west walkway connection	53	东西向的人行通道转接处
54	Walkway connection to surface parking	54	与地面停车场相连接的人行通道

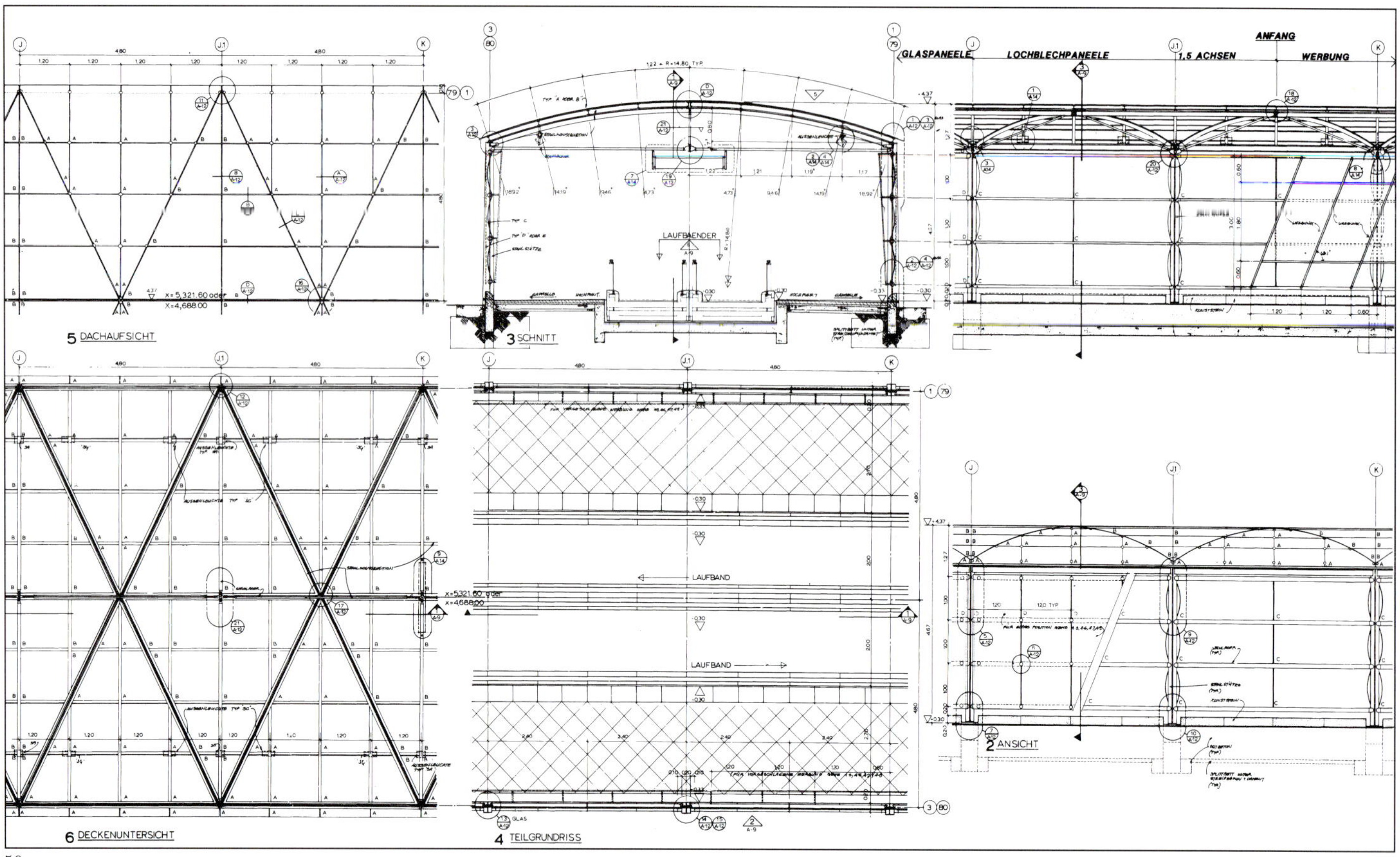

50

51

52

53

54

Hotel Kempinski/Munich Airport Center/Walkways 189

Dearborn Station

Design/Completion 1994/1997
Dearborn, Michigan
Union Station Company
80,000 square feet
Lightweight steel cable trusses
Exterior membrane of Teflon-coated fiberglass alternating with clear vision glass

迪尔伯恩火车站

设计/竣工　1994/1997
密歇根州，迪尔伯恩市
联合车站公司
80,000平方英尺
轻质钢缆桁架
特氟纶涂层玻璃纤维外墙薄膜与
透明观景玻璃交错处理的外观

The new Amtrak station to be located in Dearborn, Michigan, will be a state-of-the-art transportation facility, specifically designed for the high-speed rail system that will ultimately link Detroit with Chicago, St Louis, Milwaukee, and beyond. The station will simultaneously serve as an exhibition space for Greenfield Village, an exhibition complex containing one of the world's largest collections of transportation technology.

The design of the 900-foot-long station is based on a cable-truss structural system which supports an enclosure of fabric and glass. The elliptical form totally envelopes the platform/exhibition space, as well as the passenger cars and locomotive of the high-speed train. Access to the elevated platform is via the Amtrak ticket station at the west end, and via the Greenfield Village entrance to the east.

这个新建的全国铁路客运公司车站位于密歇根州的迪尔伯恩市，它将成为一个富有艺术表现力的交通设施，是专门为一个高速铁路交通系统设计的，该系统将最终把迪尔伯恩与芝加哥、圣路易斯、密尔沃基及更远的地区连接在一起。该车站将同时作为格林菲尔德村的一个展览空间，它将成为一个拥有世界上在交通运输技术方面的收藏品最多的一个综合性展览建筑。

这个长达900英尺的车站在设计上采用了一个钢缆桁架结构系统，由它来支撑一个用织物和玻璃做的围合体。其椭圆形的外观将站台/展览空间以及乘客车厢和高速火车的牵引机车都完全包裹起来。旅客需要穿过位于西端的全国铁路客运公司的售票站，然后再穿过东端的格林弗尔德村的入口才能到达架高起来的站台。

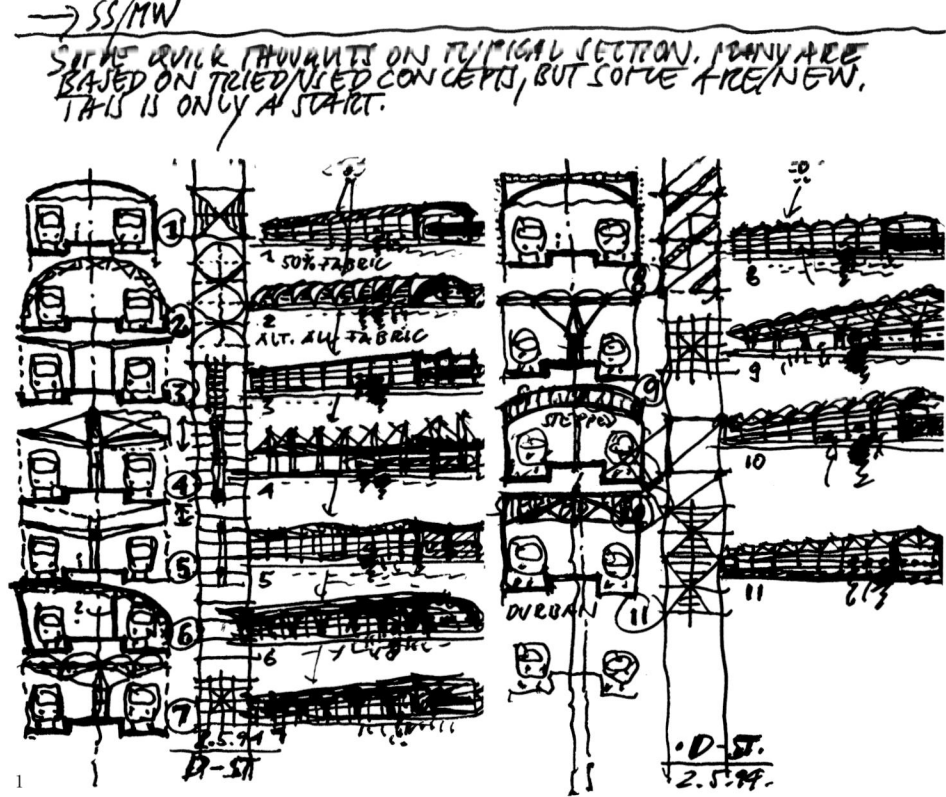

1

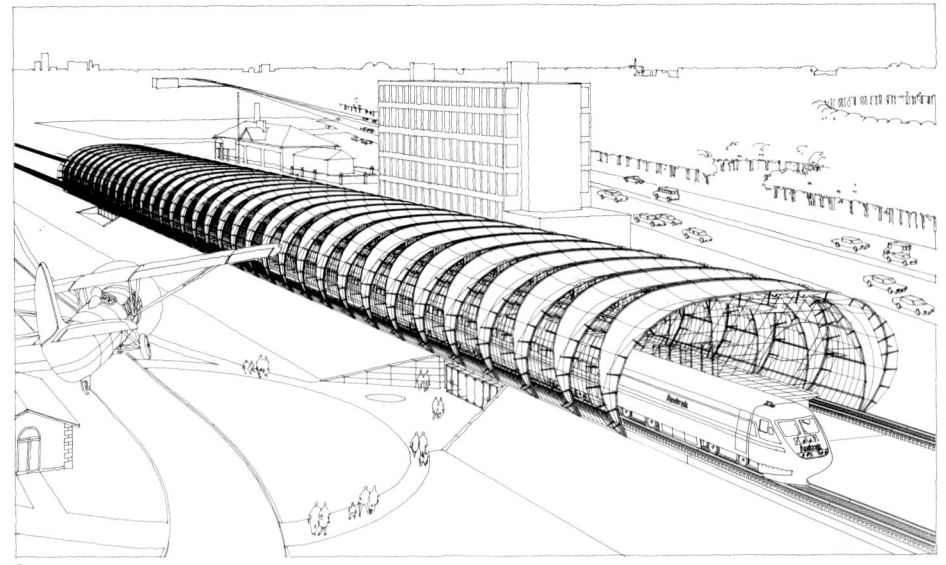

2

1	Concept sketches		1	构思草图
2	The 900-foot-long station		2	900英尺长的车站
3	Station interior		3	车站室内
4	Station exterior		4	车站室外
5	Station interior		5	车站室内

3
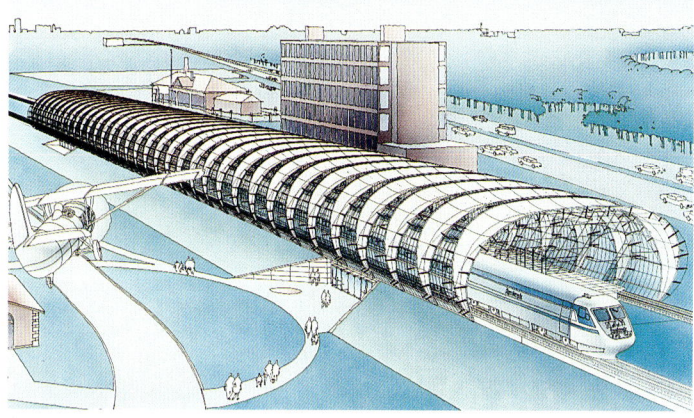
4

5

Dearborn Station 191

Bangkok International Airport

Design/Completion 1994/2001
Bangkok, Thailand
Airports Authority of Thailand
4,500,000 square feet
Steel frame with long-span trusses; concourses are three hinged, cable-stayed arches
Glass and metal panel curtain wall

曼谷国际机场

设计/竣工　　1994/2001
泰国，曼谷市
泰国机场管理局
4,500,000平方英尺
大跨度桁架钢结构：候机大厅采用三铰形钢缆固定券拱
玻璃与金属板幕墙

The Second Bangkok International Airport will be a phased development located on a vacant site outside of Bangkok. Murphy/Jahn submitted three schemes in the competition.

Concept A
This concept emphasizes passenger circulation over aircraft circulation. Because the curbfront area is not capable of providing for the ultimate capacity of the site, two separate entrances were necessary; however, this is still the most compact terminal of those studied.
A large roof trellis structure placed over the complex of functionally separate buildings unifies the site and provides an overriding consistent architectural image. Sized to accommodate future growth, the trellis provides an important functional advantage as well as an architectural one.
Continued

曼谷第二国际机场位于曼谷市郊的一块空置的场地上，将进行分期开发建设。墨菲/扬建筑事务所在这一竞赛中提交了三个设计方案。

设计构思A
　　该设计构思对乘客流线的强调胜过飞机的流线。由于道路边上的区域无法满足该场地的最大乘客容量的需要，因此需要分开设置两个入口；尽管如此，该机场航站楼的设计与其他被研究的设计相比较，它仍然是布置最紧凑的。巨大的屋顶格架结构位于这个各功能被分隔开来的综合性建筑之上，将整个基地统一起来，并形成了最为突出的连贯一致的建筑形象。为了使机场可容纳在今后不断增长的乘客数量，这个格架结构提供了一种重要的功能上的有利条件，也为建筑艺术处理带来了方便。

（待续）

1　设计构思A：设计草图
2　设计构思A：鸟瞰图/平面图
3　设计构思A：屋顶格架、售票区和管状候机大厅
4　设计构思A：屋顶格架、建筑结构

1　Concept A: design sketch
2　Concept A: aerial view/plan
3　Concept A: roof trellis, ticketing area and tubular concourses
4　Concept A: roof trellis structure

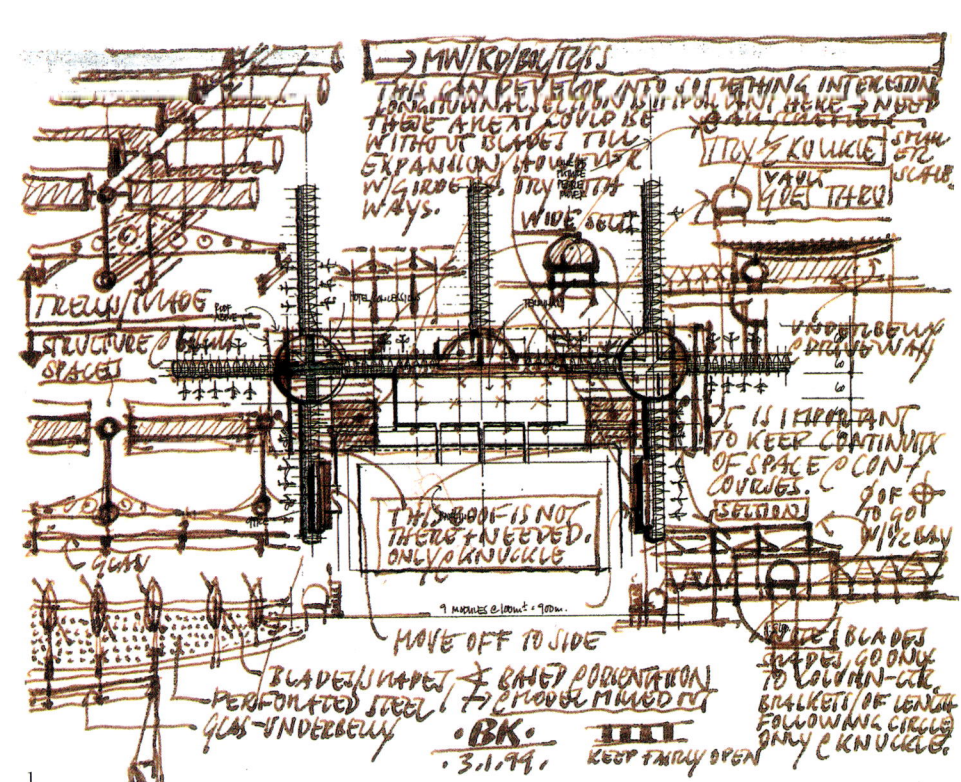
1

2

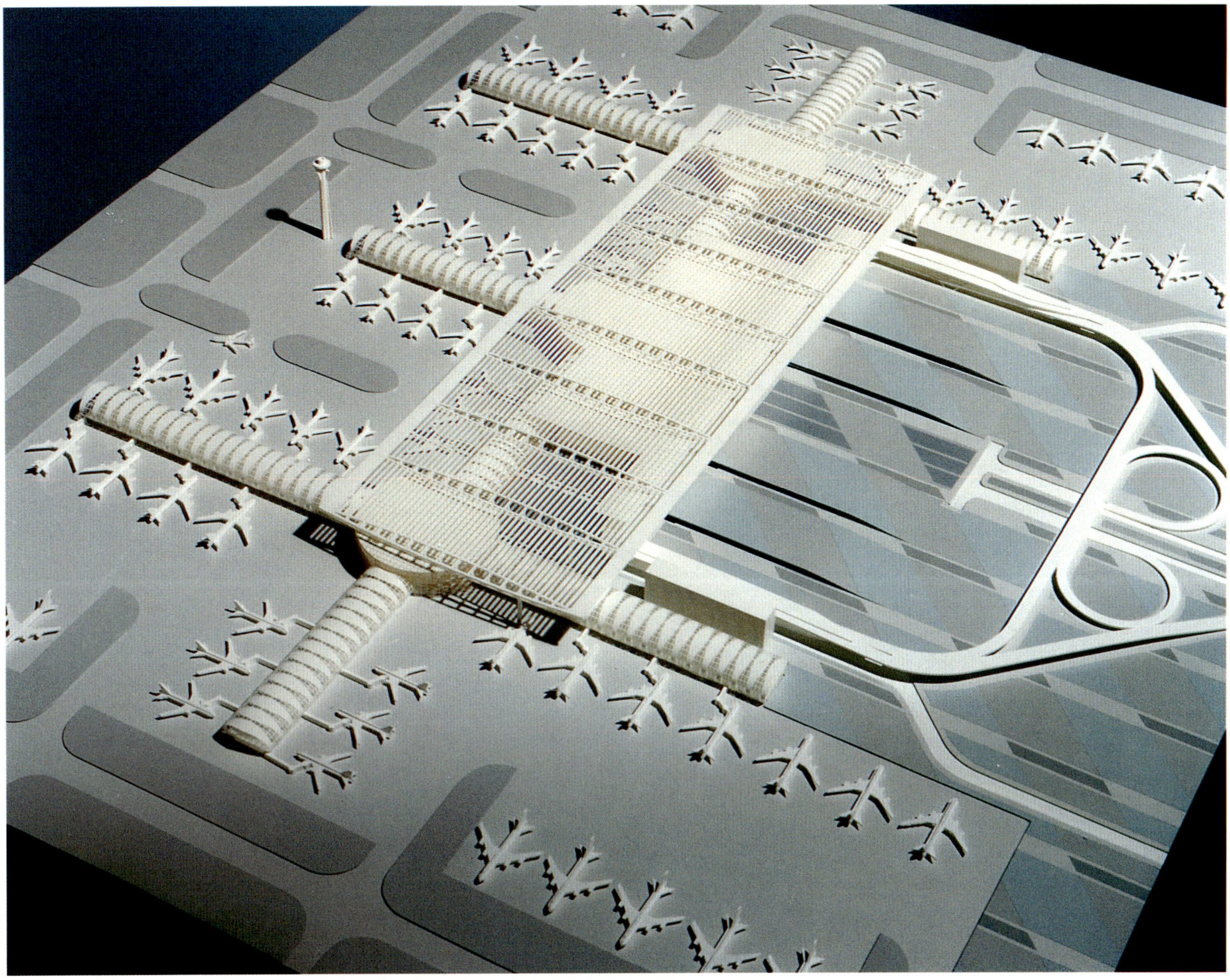

3

4

Bangkok International Airport 193

Concept B-1

By increasing the aircraft maneuvering area between concourses, Concept B-1 balances airside circulation and passenger circulation. It can be expanded so that its curbfront can accommodate more than 60 million passengers without the need for a second airport entrance to a second terminal complex.

Its architecture is based on a unit structural module of triangular domed segments. Each segment has an outer roof shell, an inner airspace containing the structural members, and an inside ceiling surface. Triangular skylights are located on the outer surface.

Concept B-2

Because of its functional configuration, Concept B-2 provides the greatest airside efficiency of all the concepts studied. Like Concept B-1, this concept could be adapted to either a single or double-entrance master plan. It requires, however, an automated people-mover in the first phase.

5

6

设计构思 B－1

通过增加候机大厅之间的飞机机动区的面积，设计构思 B－1 使得飞机专用一侧的流线与乘客流线之间形成了平衡。该方案具有可扩展性，使得其临道路边上的区域可以在不需增加第二个通往第二航站综合楼的入口的情况下，年乘客流量就可以超过 6000 万。

该建筑在艺术处理上以三角形的部分穹顶所形成的单元式结构模块为基础。每一穹顶部分都有一个外层的屋顶壳体、一个内层的包含了结构构件的空气层以及一个在内部的顶棚表层。三角形的天窗位于外层的表面部分。

设计构思 B－2

由于该航站楼的功能布置，在所有研究的构思方案中，设计构思 B－2 的飞机起降区域的效率是最高的。与设计构思 B－1 相似，这一设计构思既可采用单入口也可采用双入口的总体规划方式。但是，在建设的第一个阶段，该方案需要设置一个自动乘客传送带。

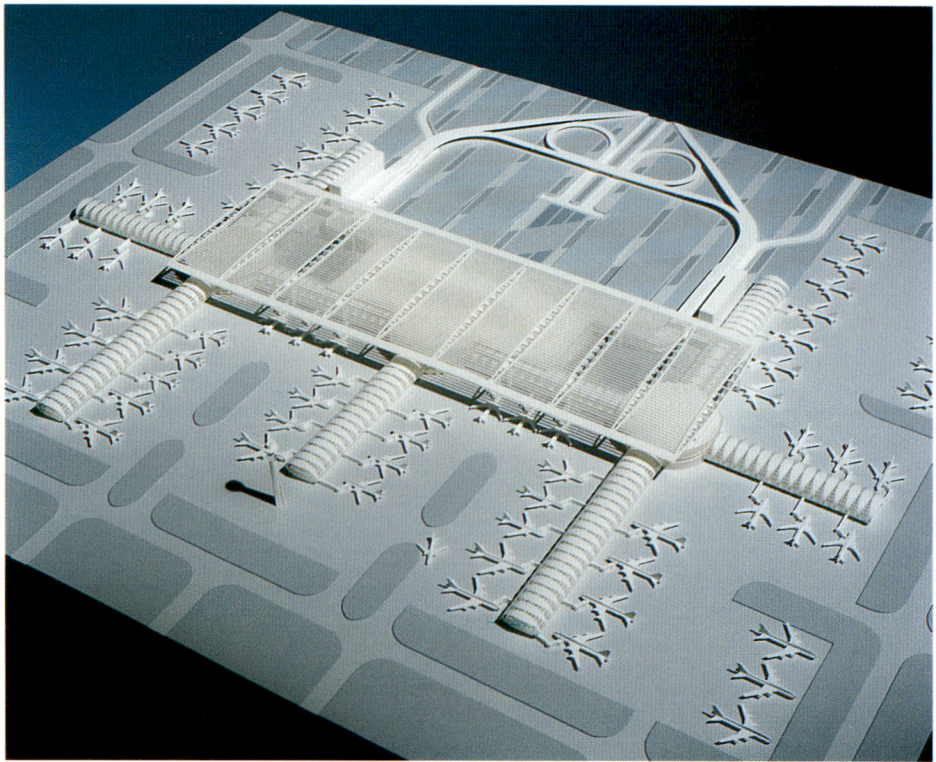
7

5　Concept A: ticketing pavilion
6　Concept A: tubular concourse
7　Concept A: airside view
8–9　Concept B-1: triangular dome segments/courtyard
10–11　Concept B-2: double cantilevered folded plate truss

5　设计构思 A：售票中心
6　设计构思 A：管状候机大厅
7　设计构思 A：飞机起降区域鸟瞰
8–9　设计构思 B-1：三角形的部分穹顶/庭院
10–11　设计构思 B-2：双层悬臂式折叠板桁架

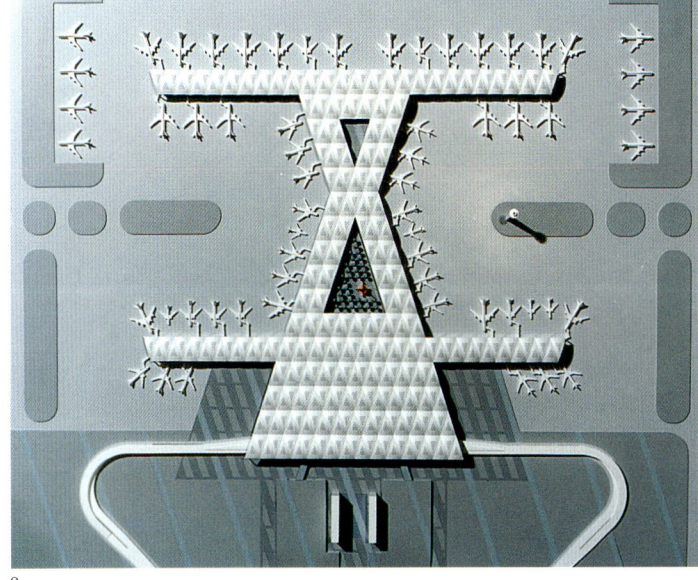
8

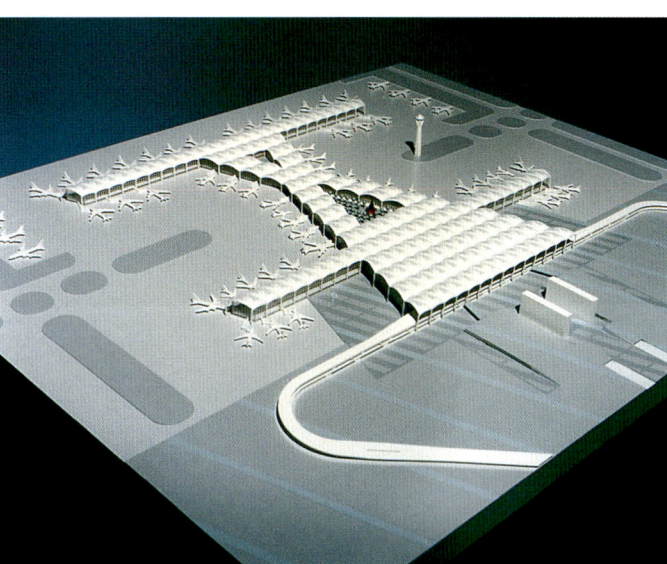
9

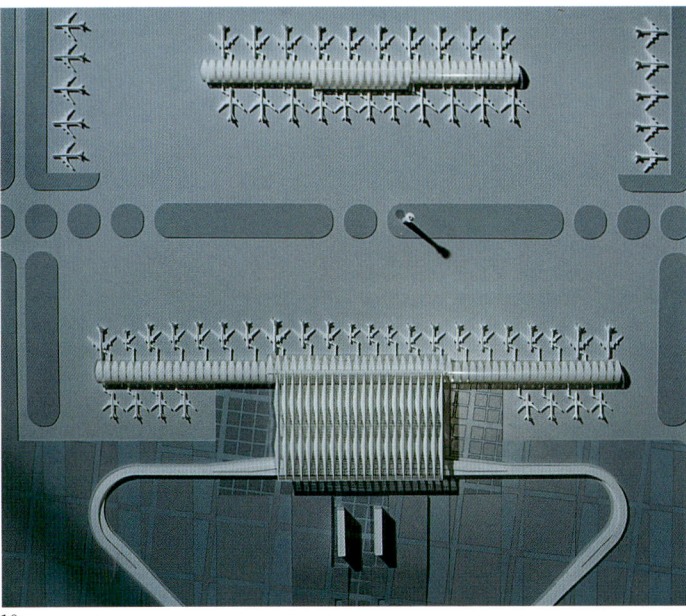
10

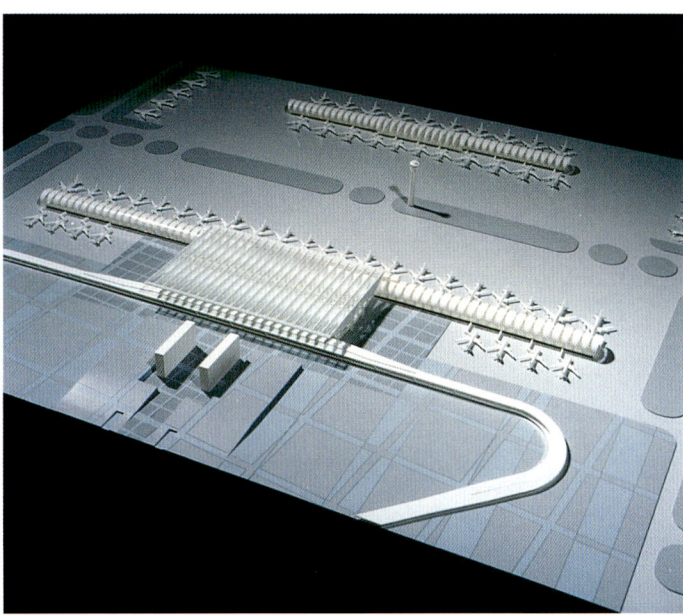
11

Bangkok International Airport　195

Flughafen Köln/Bonn Terminal 2

Design/Completion 1992/2000
Cologne, Germany
Flughafen Köln/Bonn GmbH
5,000,000 square feet
Steel roof and columns placed on concrete frame substructure
Aluminum and glass curtain wall

科隆/波恩机场第二航站楼

设计/竣工　1992/2000
德国，科隆市
科隆/波恩机场股份有限公司
5,000,000平方英尺
放置于混凝土框架基础上的钢基础和柱子
铝材和玻璃幕墙

The existing airport is a landmark in the landscape. A reconfiguration is required to accommodate passenger growth and because new plane types require new functional and circulation areas as well as new security standards and updated baggage handling. The building program begins with the construction of the new terminal, Concourse D, Parking Structure North and the renovation of the existing terminal. This phase more than doubles the existing passenger volume by the year 2000.

The expansion extends the horizontal forms of the existing terminal, using lightweight steel rather than concrete. Similar to the existing terminal, the new buildings employ a repetitive long-span structure, establishing continuity in the scale and spirit of structural expression.

The construction consists of steel piers supporting a steel roof structure with diagonal skylights. This kind of repetitive "systems building" is ideal for phased construction, provides operational flexibility, and can be expanded in a very natural and rational sequence.

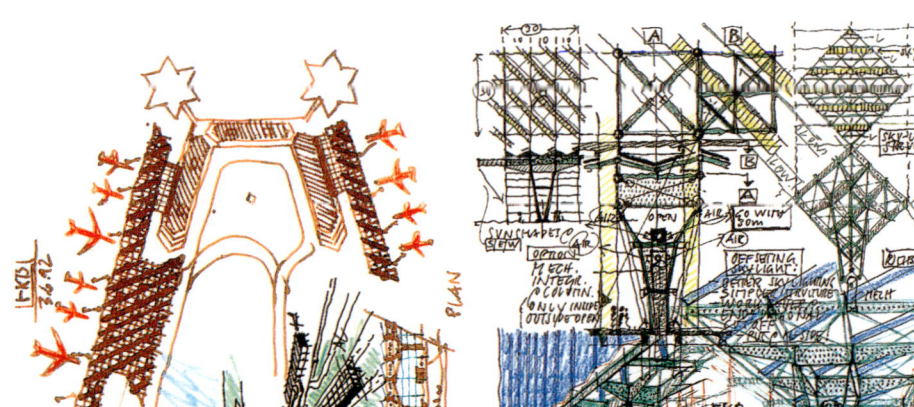

1

2

现有的机场是这一地区的一个标志性建筑物。为了容纳更多的乘客，也因为新的机型要求机场有新的功能区和流线区，以及新的安全标准和最新的行李托运服务，该机场需要进行重建。该建筑计划开始于新航站楼、D候机大厅、北侧的停车库的建设以及对现有的航站楼改造。该阶段工程可以到2000年使现有的旅客数量增加一倍多。

对现有航站楼的扩建将延伸其水平的形式，在设计上采用了轻型钢材而不是混凝土。与现有的航站楼相似，新建筑也采用了一种重复出现的大跨度结构，在结构表现的尺度和精神特征上形成了连续感。

该结构采用钢支柱来支撑一个带有成对角线的天窗的屋面结构。这种反复出现的"系统化建筑"对分期建设的项目非常理想，具有在操作上的灵活性，也可以按照一种非常自然和理性的序列进行扩展延伸。

1　Design sketches
2　Landside/terminal/parking facility, phase II
3　Existing terminal with new terminal, phase I
4　Airside view, phase I

1　设计草图
2　二期工程：停机坪/航站楼/停车场设施
3　一期工程：现有的航站楼与新航站楼
4　一期工程：停机坪一侧的模型立面

3

4

Flughafen Köln/Bonn Terminal 2 197

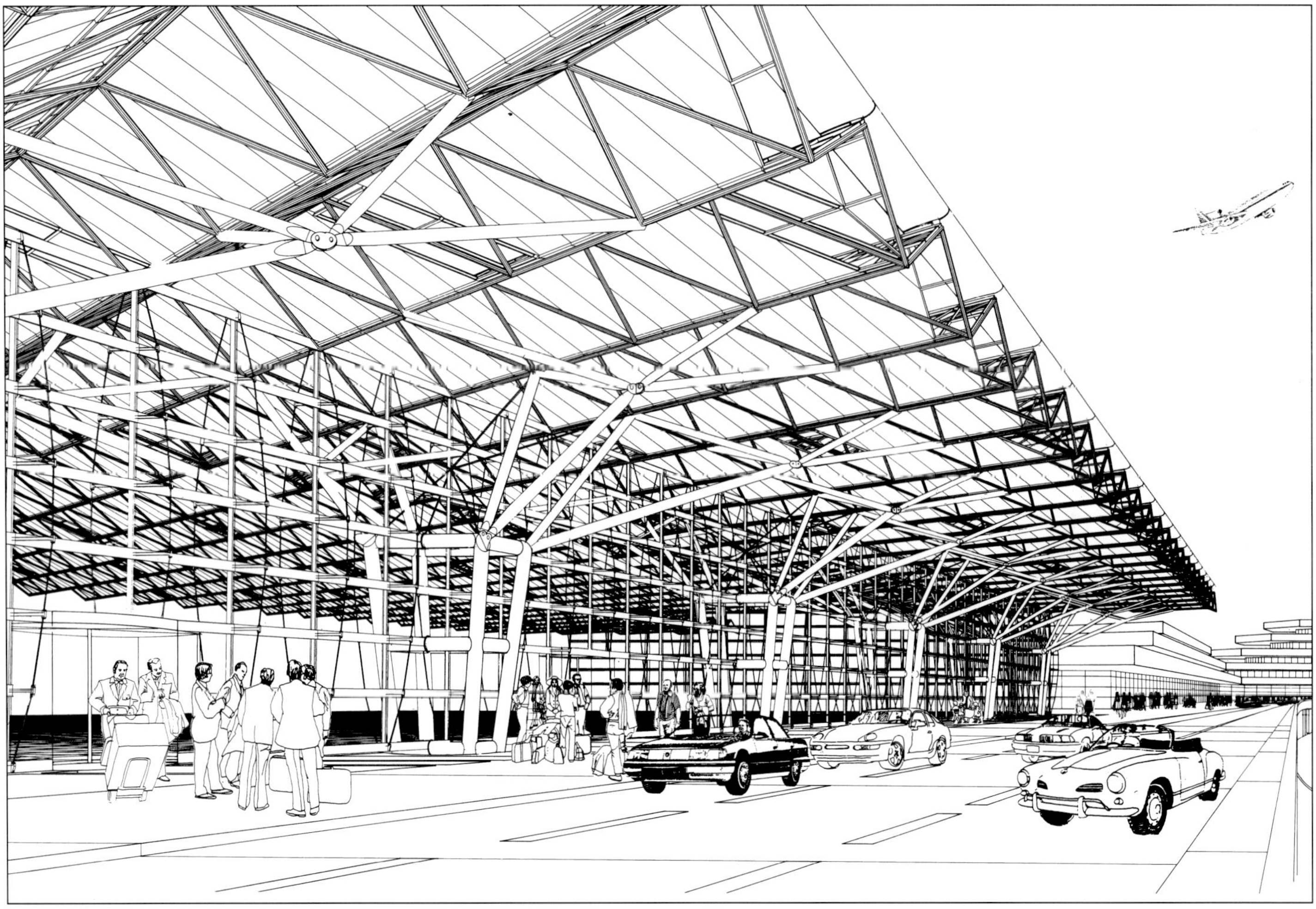

5

6

5	Covered drop-off
6	Meeters/greeters lobby
7	Bi-level roadway serving arrivals and departures
8	Ticketing/departure hall

5	有屋面的停车区
6	迎宾大厅
7	用于到达和出发的双层车道
8	售票大厅/候机大厅

7

8

Flughafen Köln/Bonn Terminal 2

Hyatt Regency Roissy

Design/Completion 1988/1992
Paris, France
Hyatt International Corporation; Toa Investment Pte. Ltd
Associate Architect: Arte J.M. Charpentier
280,000 square feet (400 rooms)
Concrete structure and steel roof structure
Painted aluminum panels and mullion system with fritted and clear vision glass on a tile base
Painted steel and fritted glass roof structure and end-walls

戴高乐机场凯悦大酒店

设计/竣工　1988/1992
法国，巴黎市
凯悦国际集团公司，托阿投资信托有限公司
合作建筑设计单位：阿特·J·M·夏邦蒂埃尔事务所
280,000平方英尺（400个房间）
混凝土结构和钢屋面结构
上漆的铝板和竖框系统及在面砖层上的热熔和透明观景玻璃窗
上漆的钢和热熔玻璃屋面结构及山墙面

The Hyatt Regency is a 400-room airport hotel for short-term stays, facilitating travel in and out of the Charles de Gaulle Airport. It includes meeting facilities, bistro cafe, restaurant and health club facilities.

The design is a modern interpretation of the traditional Parisian courtyard hotel. The central axis of the project is defined by two five-story hotel room blocks which form the edges for the exterior courtyard garden and the interior atrium garden. The exterior garden is further defined by a curving garden trellis wall executed in steel and glass. This central axis is terminated by the conference facility which is situated between the two hotel wings.

Two double-loaded linear hotel room wings form the basis of the plan. The top three floors of these wings contain the required 400 rooms. The lower two floors contain the front-of-house and support spaces for the interior garden atrium, bistro cafe, lobby bar, and ballroom/meeting room facility.

凯悦大酒店是一家拥有400个房间的机场酒店，可为进出戴高乐机场的旅客提供短期逗留的服务。在该建筑中还包括了会议设施、小咖啡厅、餐厅及健身俱乐部。

该建筑在设计是对传统的巴黎庭院式旅馆的一种现代式的诠释。两个5层楼高的旅馆客房建筑物限定了该项目的中轴线，这两栋建筑构成了外部庭院花园和室内中庭花园的边界。外部花园由曲线形的用玻璃和钢做成的格架花园外墙予以进一步的限定。中轴线在位于两个酒店侧翼之间的会议中心处结束。

两栋设有内走廊的酒店客房的直线式建筑构成了该方案的基础。这两栋建筑的顶上3层中设置了所要求的400个房间。在下面的两层中设置了前台服务区和室内花园中庭的辅助空间、小咖啡厅、大堂酒吧以及舞厅和会议厅设施。

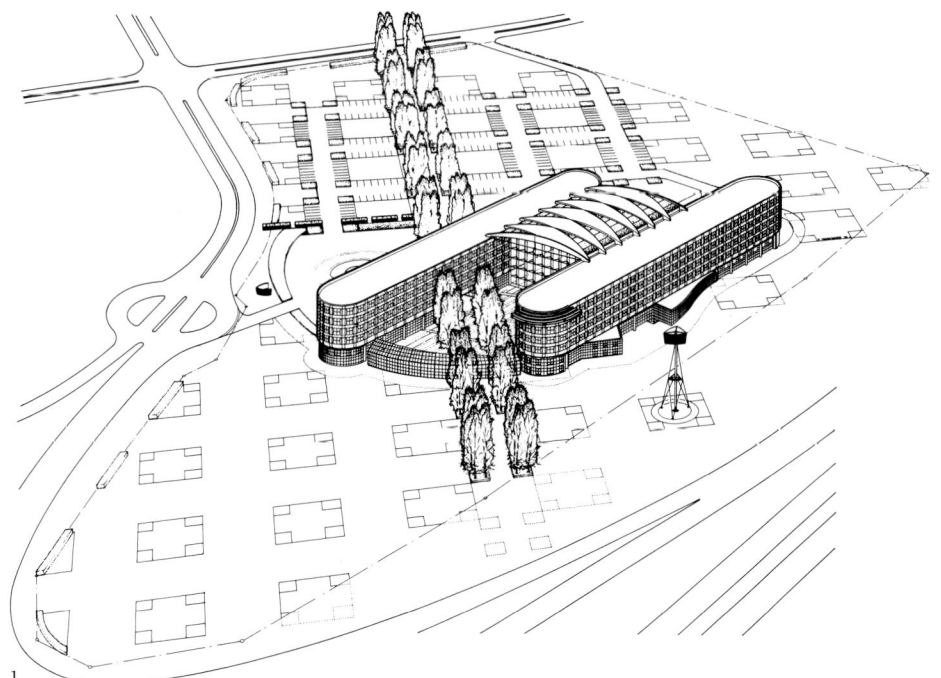

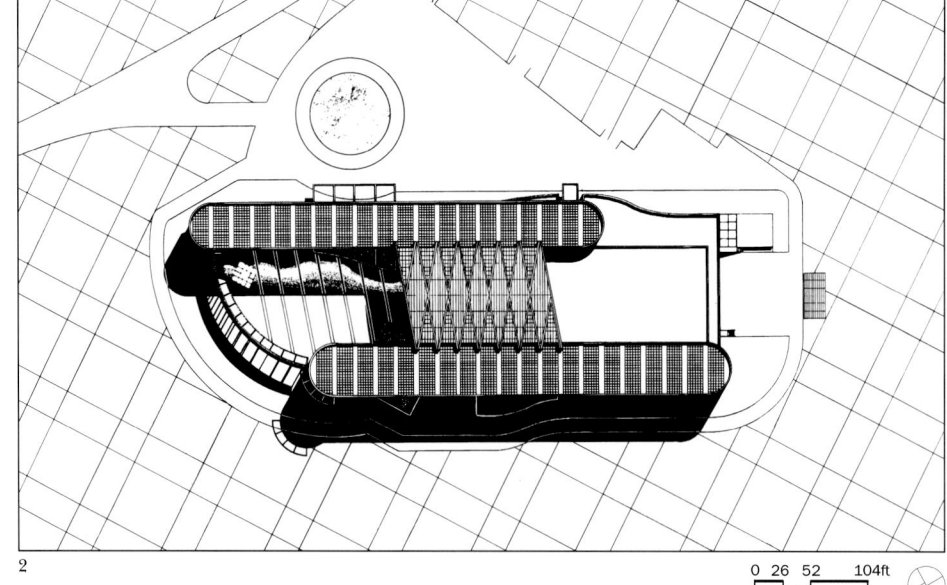

1　Axonometric
2　Site plan
3　Two five-story hotel room blocks with central atrium
4　Outdoor landscaped court with windscreen

1　轴测图
2　总平面图
3　两栋5层高的酒店客房建筑与中庭
4　室外绿化庭院与挡风玻璃

Hyatt Regency Roissy 201

5

6

7

5　Landscaped court
6　Entrance
7　Courtyard from arrival arcade
8　Courtyard at night
9　Atrium seating
10　Restaurant enclosure walls in atrium
11　Atrium interior

5　绿化庭院
6　入口
7　从到达廊道看庭院
8　庭院夜景
9　中庭里设置的座位
10　中庭中的餐厅围合隔墙
11　中庭的内部景色

8

9

10

11

Hyatt Regency Roissy　203

King Abdulaziz International Airport Competition

Design 1991
Jeddah, Saudi Arabia
International Airport Projects, Ministry of Defense and Aviation
9,000,000 square feet
Steel frame with long-span trusses infilled with metal panel and tinted colored glass
Stainless steel and glass curtain wall

阿卜杜勒阿齐兹国王国际机场设计竞赛

设计 1991
沙特阿拉伯，吉达市
国际机场项目公司，防卫和航空部
9,000,000 平方英尺
大跨度桁架的钢框架结构，采用金属板和彩色玻璃填充
不锈钢和玻璃幕墙

The competition for a new dual-terminal complex for the King Abdulaziz International Airport in Jeddah provided an opportunity to push airport design to yet further limits. Its size of 90 gates equals that of Chicago's O'Hare International Airport.

A simple arch spans the whole site, unifying the two buildings and creating a central space with mosque and gardens.

Functionally, the building is organized as a terminal concourse. Between the upper departure level and the lower arrival level, an intermediate pedestrian level is inserted, connecting the park and parking garage with the terminal buildings. The concourses have separate departure and arrival levels for international flights and also allow swing-use for domestic flights. Departing and arriving are carefully orchestrated sequences of varying spatial experiences which afford a continually varying sense of place and orientation. The passenger moves through the facility in a direct, logical way and can enjoy the experience.

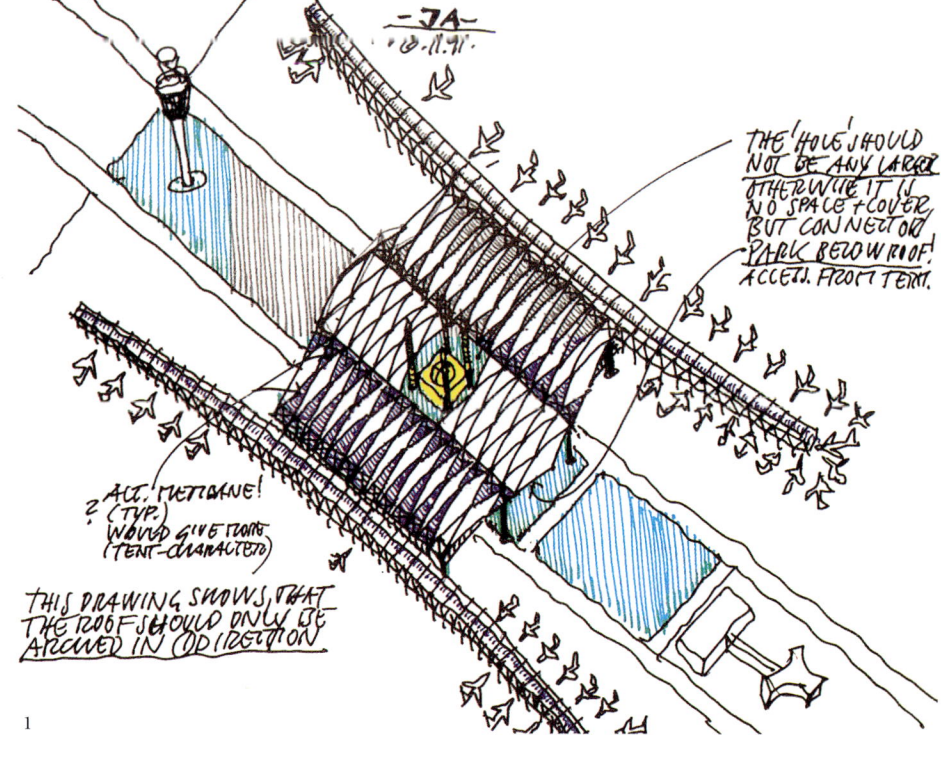
1

针对位于吉达的阿卜杜勒阿齐兹国王国际机场的新型双向航站楼综合体所举行的设计竞赛，为建筑师们提供了一个将机场设计推到更高水平的机会。这个拥有90个登机口的新机场在规模上与芝加哥市的奥黑尔国际机场相当。

一个造型简洁的拱顶覆盖着整个基地，将两栋建筑统一在一起，并形成了一个带有清真寺和花园的中央空间。在功能上，该建筑物被设计成一个航站楼的候机大厅。在上层的出发平面与下层的到达平面之间插入了一个起过渡作用的人行平面，将露天停车场和停车库与航站楼连接起来。候机大厅有用于国际航班的各自分开的出发和到达平面，也可以用于国内航班。出发和到达的路线被设计成一系列精心安排的不同空间体验，也形成了一种连续变化的场所感和方向感。旅客以一种直接的、富于逻辑的方式穿过这些设施，能够享受到这种体验的乐趣。

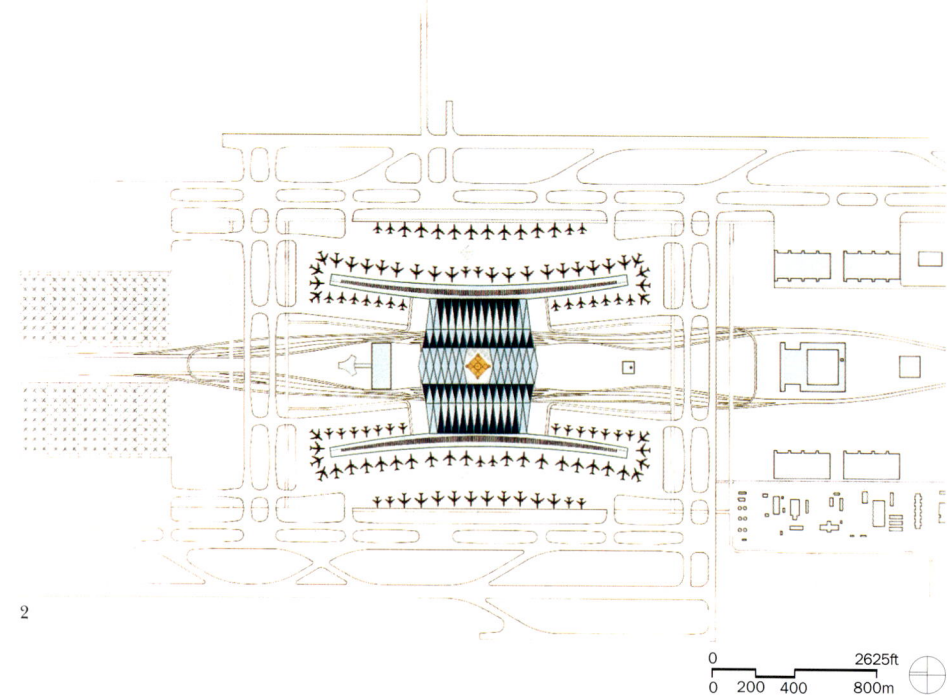
2

1　Concept sketch
2　Site plan
3　90-gate terminal facility
4　Central pavilion
5　Departures/gate area

1　构思草图
2　总平面图
3　拥有90个登机口的航站楼
4　中央大厅
5　出发区/登机口区域

3

4

5

King Abdulaziz International Airport Competition   205

阿卜杜勒阿齐兹国王国际机场概念设计

设计　　1993
4,500,000 平方英尺
模数化钢框架结构单元
不锈钢和玻璃幕墙与
降低高度的遮阳装置

King Abdulaziz International Airport Conceptual Design

Design 1993
4,500,000 square feet
Modular steel-framed structural units
Stainless steel and glass curtain wall with lowered shading devices

Following the design competition, the site was changed. The objective was to provide a clear site allowing for the phased implementation of a new terminal without disruption to any existing operations.

The design concept selected uses domed segments as the elementary structural unit. Each segment acts together with the others surrounding it to form an efficient, highly sculptural roof system. The distinctive skylight pattern alternates from the outside surface of the roof to the inner surface. As a result, all natural light is filtered and indirect.

Landscaping is an integral part of the design. A large, stepped, central courtyard brings natural light and outside views to all levels of the building.

　　在设计竞赛之后，建筑的基地发生了变化。随后的设计目标是要提供一个可以分期实施修建新航站楼的基地，不得影响任何现有的运转工作。

　　被选中的设计构思采用穹顶式的结构作为建筑物的基本结构单元。每一个片段与其周边的其他片段共同起作用，构成了一个高效而极具雕塑感的屋面系统。独具特色的天窗形式从屋面的外表面到内表面交替形成变化。正因为如此，所有的自然光线都是被过滤了的，成为间接采光。

　　环境绿化是整个设计中的一个不可分割的部分。一个巨大的、阶梯形的中央庭院将自然光和外界的景观引入到该建筑物的所有平面上来。

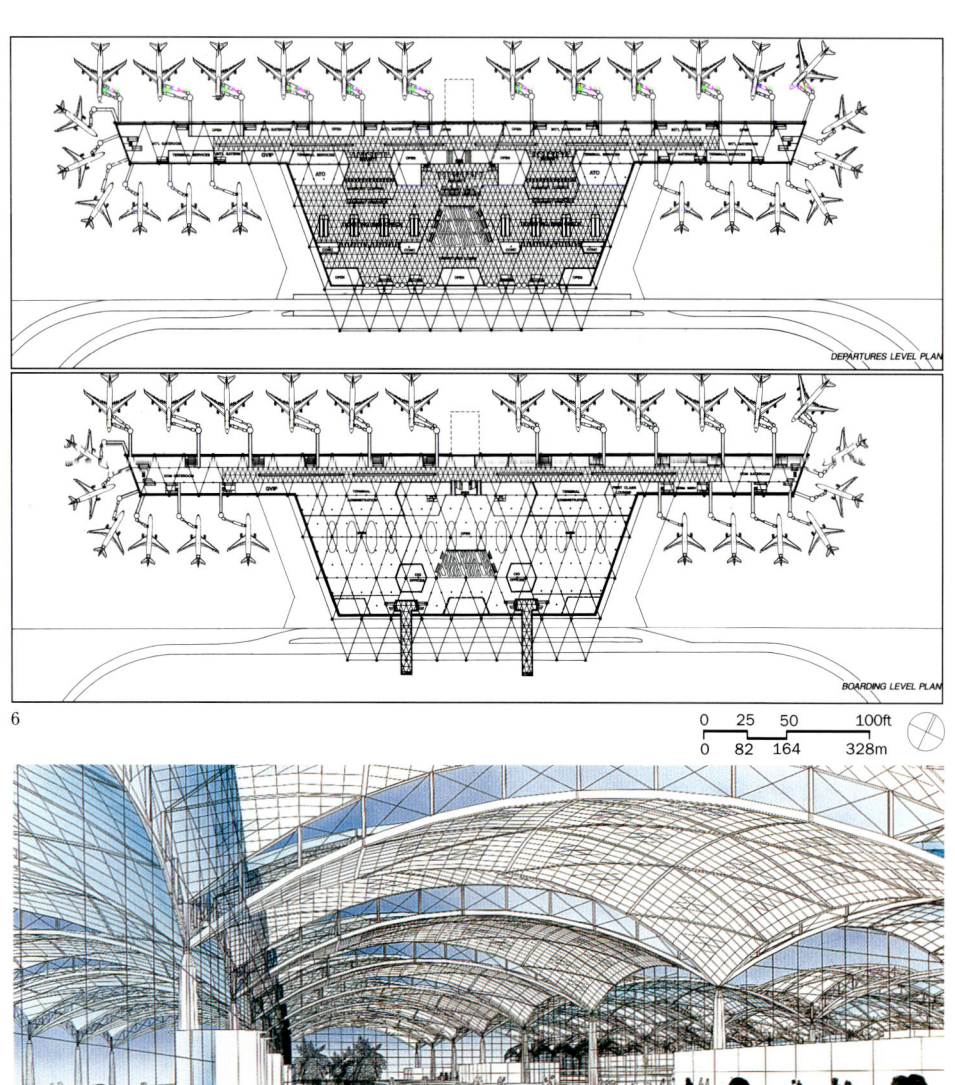

6　出发层和到达层平面图
7　由片段式的穹顶所覆盖的售票区
8　飞机停靠区方向立面，可以看到酒店客房部分的幕墙
9　从人行方向看建筑

6　Departure and arrival plans
7　Ticketing pavilion with segmented domes
8　Airside elevation showing hold room curtain wall
9　Landside view

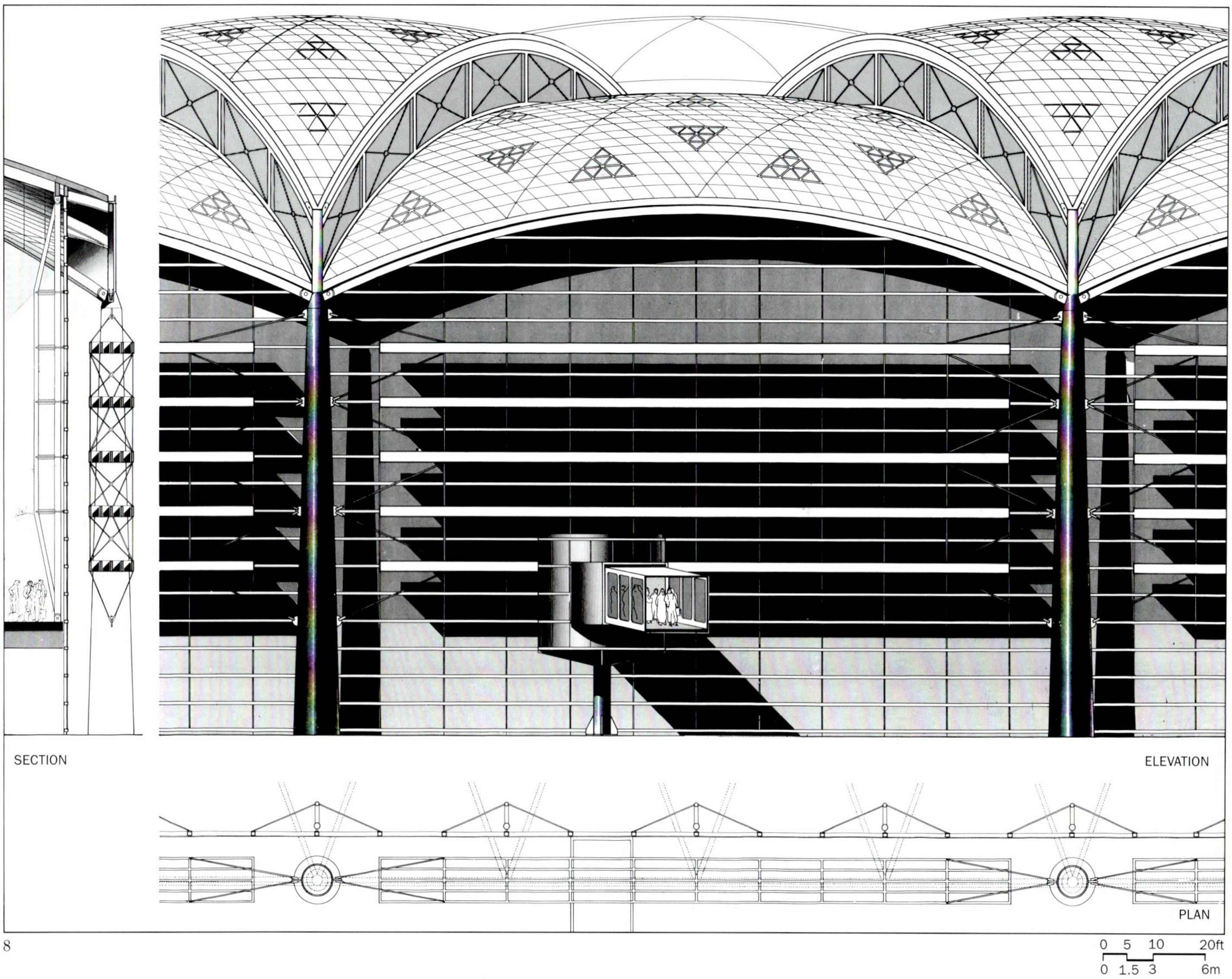

SECTION ELEVATION

PLAN

8

```
0   5   10      20ft
0  1.5  3       6m
```

9

King Abdulaziz International Airport 207

One America Plaza/Trolley Station

Design/Completion 1988/1991
San Diego, California
Shimizu Corporation
660,000 square feet
Steel perimeter columns, concrete core; exposed steel structure in station
Granite-clad columns with silicone-glazed unitized wall panels; horizontal stone spandrels with blue reflective glass infill at corners

第一美利坚广场/有轨电车车站

设计/竣工　1988/1991
加利福尼亚州，圣迭戈
清水株式会社
660,000平方英尺
周边采用钢柱，混凝土核心筒，车站部分采用暴露式钢结构
柱子采用花岗石贴面，涂硅面层的组装式外墙板，水平式的石材窗下墙，在转角处采用蓝色反射玻璃填充墙

This building is the key component in a mixed-use complex located in downtown San Diego, bordered by Broadway, Kettner, India, and B Streets. The 34-story tower fronts on Broadway, the main east–west artery of the central business district. Wrapping around the tower at plaza level is the San Diego trolley; the enclosed trolley station serves as the focal point of the two-story retail arcade which follows the serpentine form of the tracks.

The curving space of the trolley station also resolves the situation at the end of C Street, spatially linking this corridor to the plaza in front of the Santa Fe Depot. The result is an urban complex which animates the street with pedestrian activity, clarifies its relationship with the surrounding pattern of streets and open space, and incorporates a variety of public spaces.

该建筑是一个位于圣地亚哥的市中心区的多用途综合体中的最主要的组成部分，紧邻着百老汇大街、凯特纳大街、印第安大街以及B大街。这栋塔楼有34层高，正对着百老汇大街，这条街道是中央商务区中主要的东西向交通干道。圣地亚哥市的有轨电车在广场平面上环绕着这栋塔楼，封闭式的有轨电车站成为两层楼的商业零售廊道的中心，该廊道在形式上顺应着蜿蜒的电车轨道。

有轨电车车站的曲线型空间也决定了在C大街端头处的处理手法，它在空间上将这条走廊与在圣达菲火车站前面的广场连接起来。该项目实施后成为一个城市综合体，以行人的活动给街道带来了生气，明确了该建筑与周边街道和开放式空间模式之间的相互关系，并且将各式各样的公众空间组合在了一起。

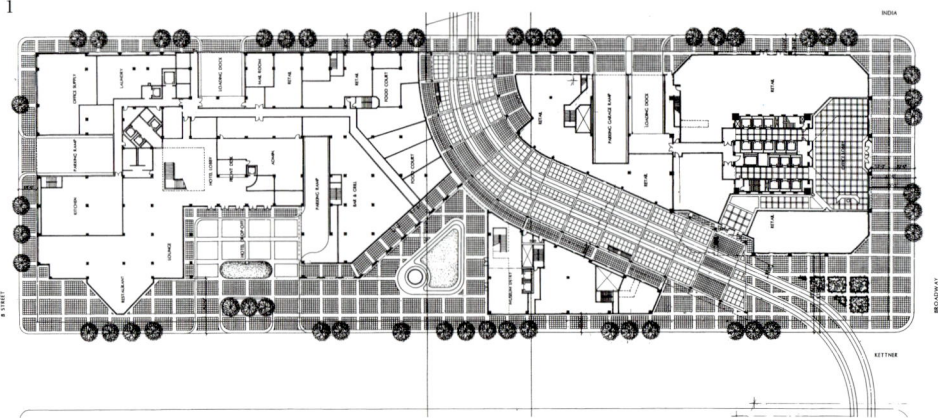

1　从印第安大街和百老汇大街方向看建筑全景
2　底层平面图
3–5　有轨电车车站

1　Overall view from India and Broadway Streets
2　Ground-floor plan
3–5　Trolley station

One America Plaza/Trolley Station 209

JFK Consolidated Terminal

Design 1988
Jamaica, New York
American Airlines/Port Authority of NY/NJ
1,800,000 square feet
Steel frame with long-span, triangulated arched trusses with metal and glass infill
Steel and glass curtain wall

J·F·肯尼迪机场联合航站楼

设计　　1988
纽约州，牙买加市
美国航空公司/纽约州与新泽西州港口管理局
180,000 平方英尺
钢框架结构，大跨度三角支撑的券式桁架，采用金属和玻璃的填充维护材料
钢和玻璃的幕墙

The terminal is configured as a four-level structure. The lower level contains all federal inspectional services areas, domestic claim areas, mechanical/electrical space and a large meeters/greeters lobby to serve the arriving passengers, their visitors and the deplaning curbfront. The grade level contains both inbound and outbound baggage systems, the PDS station and mechanical space, and ramp operations. The upper level contains ticketing, gate facilities, VIP lounges, and the enplaning curbfront. The penthouse level contains the FIS arrivals corridor and mechanical fan rooms.

The satellite concourse is configured as a three-level structure. The grade level contains ramp operations and mechanical/electrical rooms, the upper level contains gate facilities and concessions, and the penthouse level contains FIS arrival corridors and mechanical fan rooms.

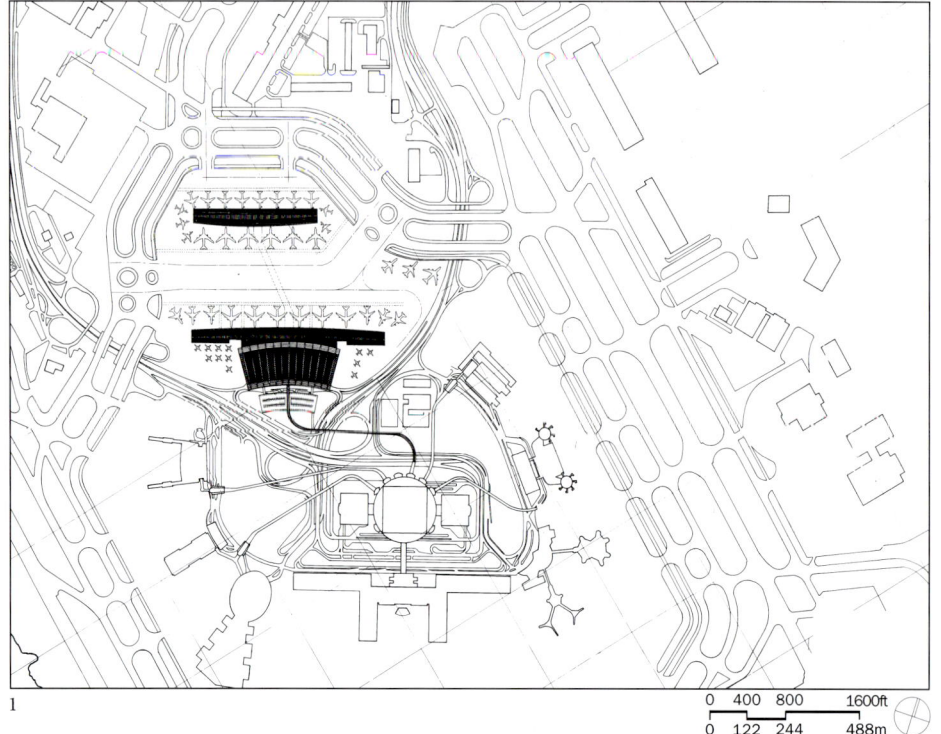

1

2

　　该航站楼被设计为一个4层楼的建筑。下面第一层中包括了所有的联邦检查服务区、国内行李提取区、机械和电气设备区、一个供到达的旅客和接机人使用的接机等候大厅以及供旅客下飞机的区域。与街面相平的首层包括进港和出港行李托运系统、PDS站和机械设备区以及引桥操作系统。上面的一层中设置了售票处、登机口设施、VIP休息室以及旅客登机区域。顶部的阁楼层中则设置了FIS到达走道和机械送气扇机房。

　　另一个卫星式的候机大厅被设计成一栋3层的建筑。地面首层中设置了引桥操作系统及机械和电气设备室，上面一层中设置了登机口设施和出租区域，而其顶层中设置了FIS到达走道和机械送气扇机房。

1　Site plan
2　Concept sketch
3-4　Terminal with landside/airside concourse

1　总平面图
2　构思草图
3-4　拥有旅客区和飞机起降区候机大厅的航站楼

JFK Consolidated Terminal 211

5 Section	5 剖面图
6 End wall	6 山墙立面
7 Hold room with train connection to satellite facility	7 与卫星式航站楼之间有列车联系的集散大厅
8–10 Ticketing pavilion	8–10 售票亭

5

6

7

JFK Consolidated Terminal 213

United Airlines Terminal One Complex

Design/Completion 1983/1987
Chicago, Illinois (O'Hare International Airport)
United Airlines
1,185,000 square feet
Folded steel truss at pavilion, perforated steel arches at concourse
External aluminum tube mullion supports; combination of clear, clear fritted, and tinted glass at vertical walls; prefabricated silicone-glazed units at vaults

联合航空公司第一航站楼

设计/竣工　1983/1987
伊利诺伊州，芝加哥市（奥黑尔国际机场）
联合航空公司
1,185,000 平方英尺
大跨度的大厅采用折叠式钢桁架，候机大厅采用带孔洞的钢拱券
建筑外部采用铝管竖框支撑体，在垂直墙面上采用透明玻璃、透明热熔玻璃和彩色玻璃，在拱顶部分采用预制硅树脂透光单元

The Terminal One Complex for United Airlines at Chicago's O'Hare International Airport provides 48 gates, 12 commuter parking positions and 1,400,000 square feet of operating facilities.

The two primary components of the project, Concourses B and C, form two linear structures with a length of 1,600 feet. A gap of 815 feet between them allows for dual taxiing of wide-bodied aircraft. Concourse C contains 30 gates, and the remaining 18 are in Concourse B. A commuter airline operation is housed in a related wing adjacent to Concourse B.

The planning of the complex is based on the concept of parallel concourses, a departure from the Y-shaped concourses prevalent at O'Hare. The parallel configuration eliminates the "dead-ends" created in the Y-shaped form. This reduces aircraft waiting time on the apron, allows two-way taxiing which expedites aircraft movement to the gates, and makes the total perimeter of Concourse C available for aircraft parking.

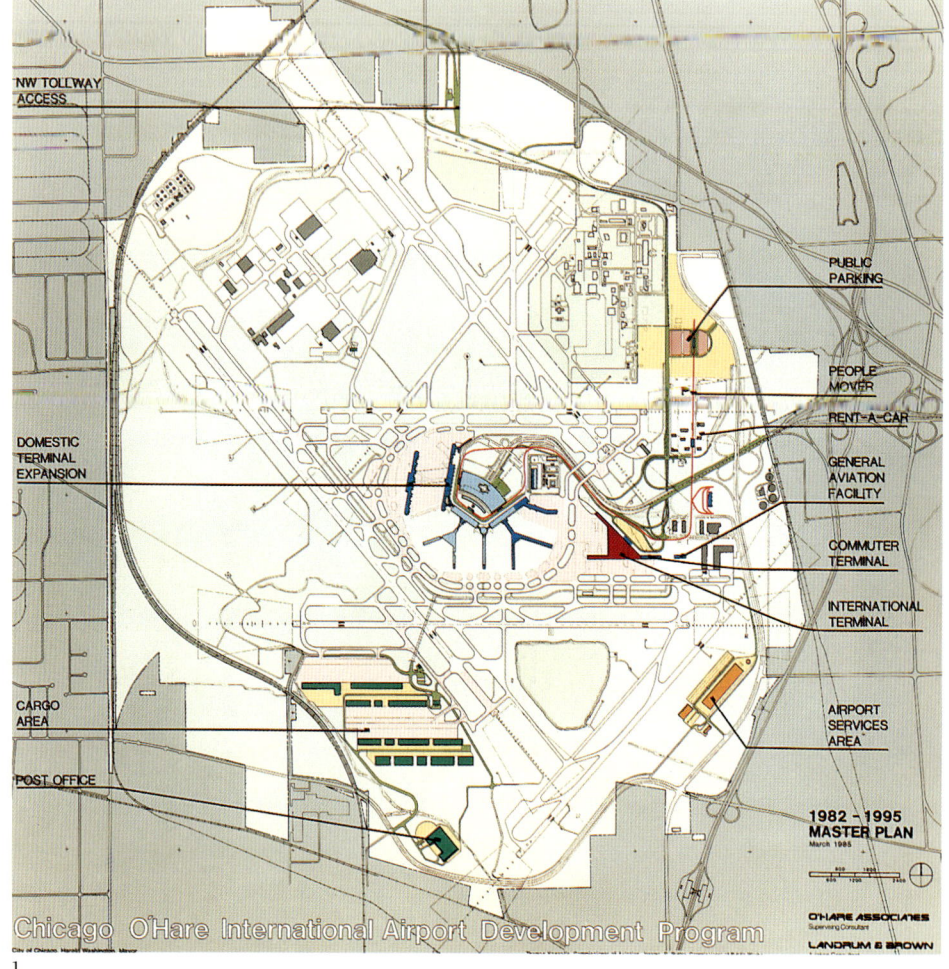

在芝加哥奥黑尔国际机场为联合航空公司服务的第一航站综合楼，设计了48个登机口、12个通勤车停车位，以及1,400,000平方英尺的运作设施。

该项目的两个主要的构成部分是候机大厅B和候机大厅C，这两栋建筑形成了长度为1600英尺的直线性结构。在其间的一个815英尺宽的缝隙可以满足宽体飞机双向滑行的要求。候机大厅C设置了30个登记口，其余的18个在候机大厅B里。与候机大厅B相接的一个侧翼专门用于一个通勤航线的运营。

这一综合性建筑的规划是建立在平行布置候机大厅的构思之上的，是从在奥黑尔机场上流行的"Y"型候机大厅布置方式上演化出来的一种形式。平行布置所形成的构图消除了在"Y"型结构上所形成的"死尽端"。这种设计减少了飞机在停机坪上的等待时间，可以实现飞机双向滑行，使飞机可以快速到达登机口处，并使候机大厅C的整个周边范围都可以供飞机停靠。

1　Site plan
2　Landside Concourse B: departure drop-off
3　Sections
4　Concourse B: airside view
5　Overall view of Concourses B and C

1　总平面图
2　候机大厅B的地勤区：出发下车处
3　剖面图
4　候机大厅B：空勤区全景
5　候机大厅B和C的全景照片

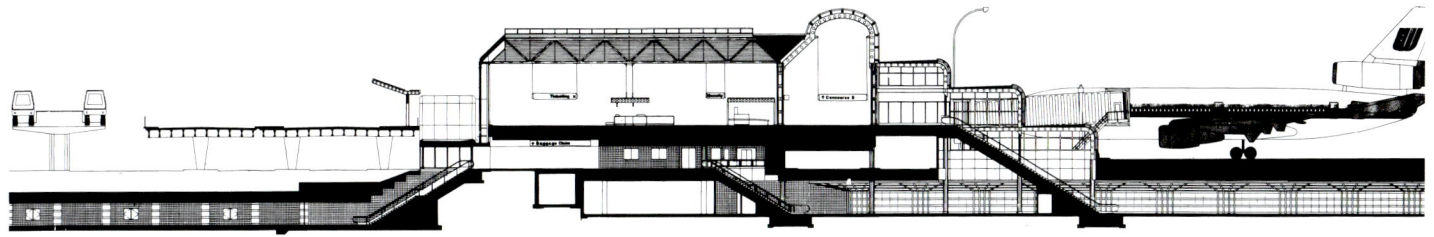
TERMINAL 1 • CONCOURSE B

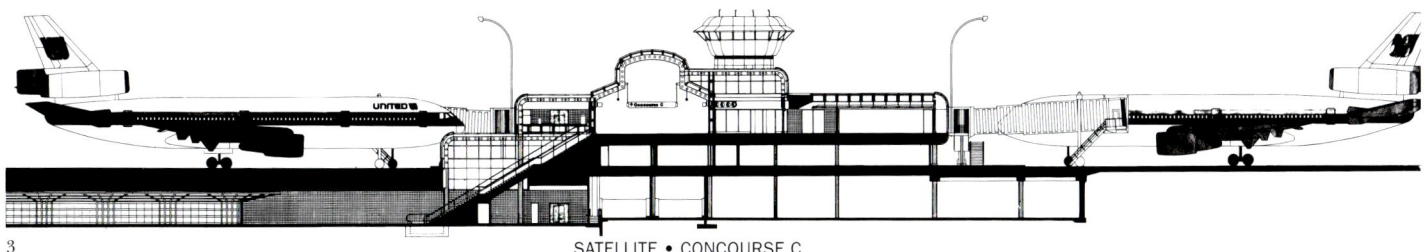
SATELLITE • CONCOURSE C

United Airlines Terminal One Complex 215

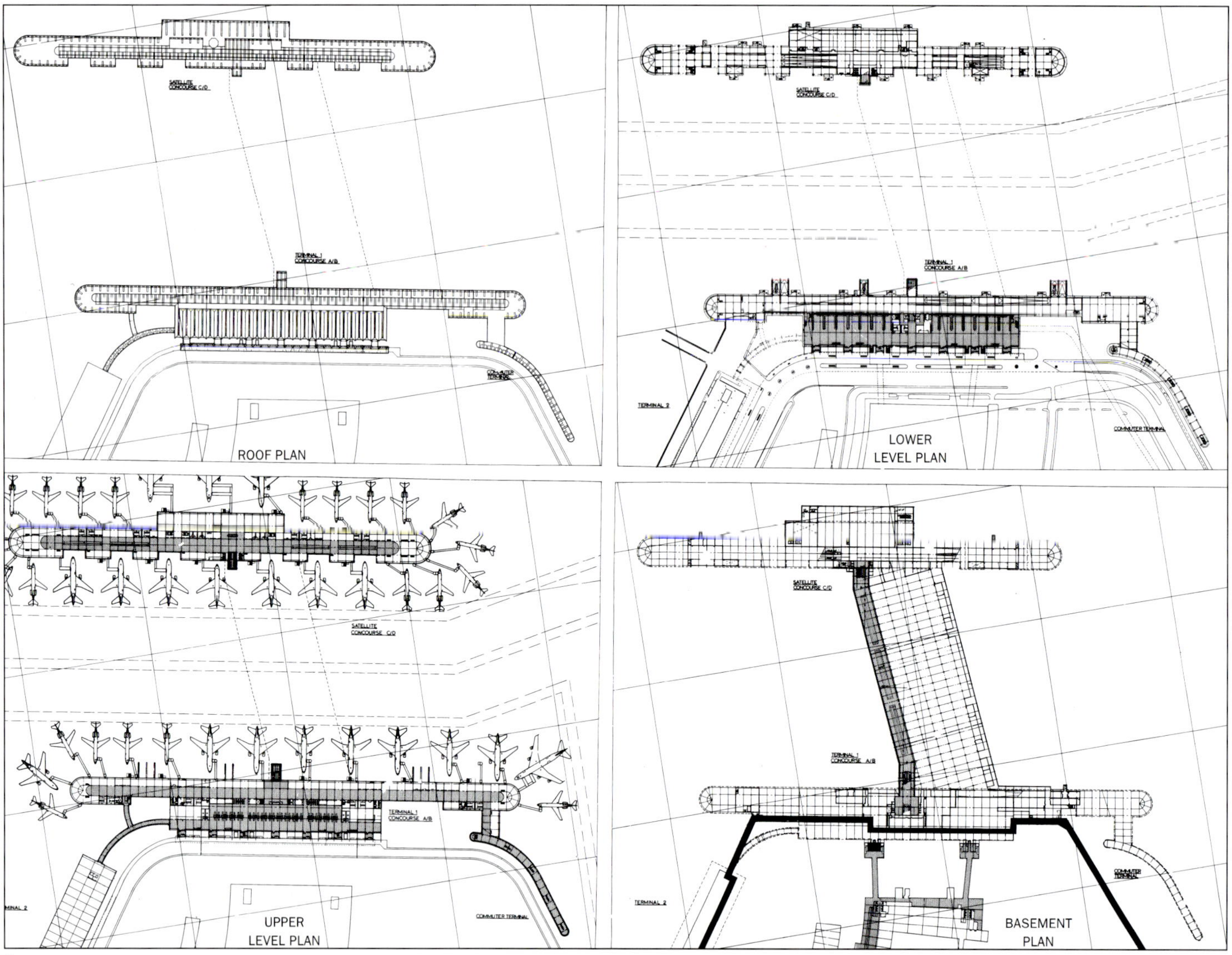

6 Concourse plans
7 Roadway
8 Curtain wall detail
9 Stair detail
10 Pedestrian tunnel access
11 Concourse B interior

6 候机大厅平面图
7 车行道
8 幕墙细部
9 楼梯细部
10 人行通道入口
11 候机大厅 B 的室内处理

8

9

10

11

12　Ticketing pavilion
13　Concourse B termination
14–15　Underground pedestrian tunnel
16　Concourse B vault

12　售票处
12　候机大厅 B 的尽头
14–15　地下人行通道
16　候机大厅 B 的拱顶内部

12

13

14

15

16

United Airlines Terminal One Complex

O'Hare Rapid Transit Station

Design/Completion 1979/1984
Chicago, Illinois
Chicago Building Commission
Reinforced post-tension concrete with sprayed concrete sloping berms
Glass block walls

奥黑尔高速交通中转站

设计/竣工　　1979/1984
伊利诺伊州，芝加哥市
芝加哥建筑委员会
后张拉法预应力钢筋混凝土与喷射式混凝土斜面护堤道
玻璃砖砌墙体

The O'Hare Rapid Transit Extension is the final link between the Chicago Loop and Chicago O'Hare International Airport. The station, which becomes the terminus of the rapid transit line, is located beneath the O'Hare Parking Structure and consists of two components: the trainroom and the pedestrian concourse.

The trainroom contains three tracks and two platforms, and is approximately 600 feet long, 70 feet wide, and 30 feet high. The pedestrian concourse, which connects the station with four underground pedestrian tunnels leading to the three airport terminals and the hotel, is approximately 1,100 feet long, 45 feet wide and 18 feet high.

To obtain a column-free platform area, the structural system developed for the station consists of large post-tensioned concrete girders which transfer the load from the parking structure columns above the trainroom. This system will be constructed using an open-cut excavation, resulting in sloping berms sprayed with concrete outside the trainroom for the entire length of the station.

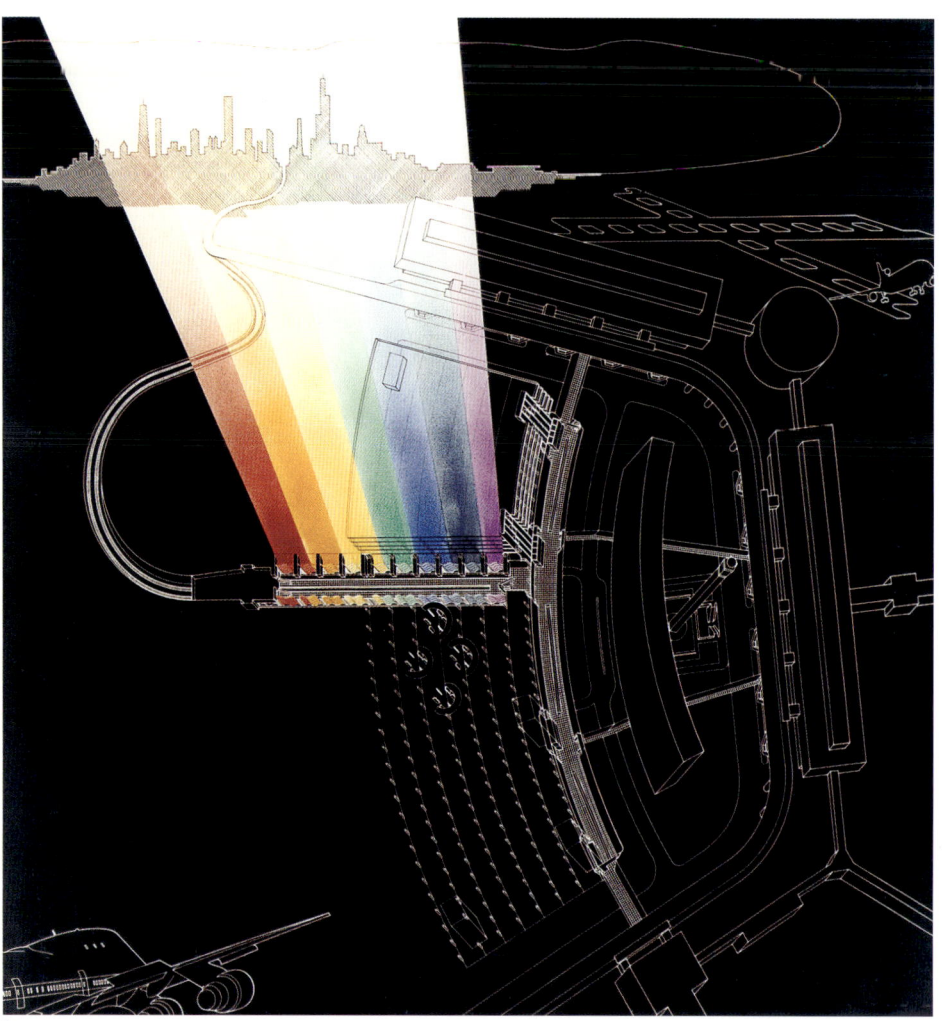

1

奥黑尔高速交通中转站扩建工程是在芝加哥的外环路和芝加哥的奥黑尔国际机场之间的重要联系纽带。作为高速交通路线的终点站，该车站位于奥黑尔停车楼的下面，由两个部分所组成：列车站台厅与旅客候车大厅。

站台厅包括三条轨道和两个站台，其长度约为600英尺，宽度为70英尺，高度为30英尺。旅客候车大厅将车站与四条地下人行隧道连接起来，这些隧道可通往三个机场航站楼和一家酒店。候车大厅长约1,100英尺，宽45英尺，高18英尺。

为了实现一个无柱的站台区，该车站采用的结构体系由巨大的后张拉混凝土大梁构成，它们将位于站台大厅上面的停车场的结构柱上传下来的荷载传递下来。该系统采用露天开挖的方式进行施工，从而在整个车站站台厅外部沿车站的整个长度形成了用喷射混凝土的方式修建的倾斜的护堤。

2

1　Site plan
2　Three-track platfrom
3　Ticketing area
4　Train platform
5　Train platform with glass block wall

1　总平面图
2　有三条轨道的站台区
3　售票区
4　列车站台
5　采用玻璃砖砌筑墙体的列车站台

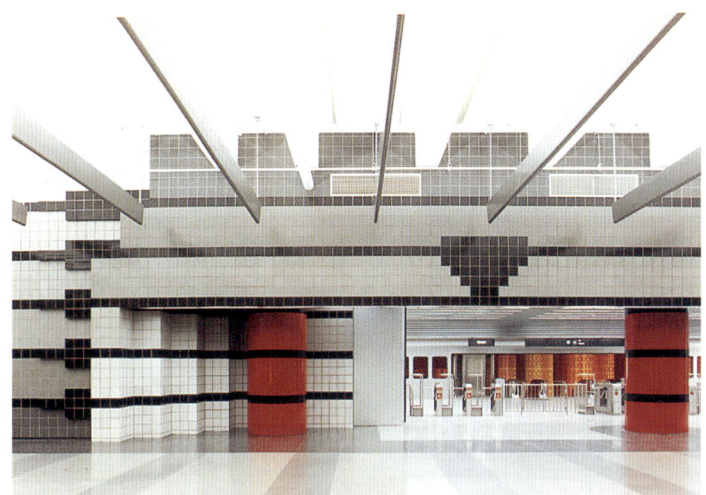
3

4

O'Hare Rapid Transit Station

Firm Profile

事务所简介

Office of Murphy/Jahn, Architects

The office of Murphy/Jahn, Architects is located on one full and two partial floors. The design is based on classical principles of space organization for an architect's office: open design studios for free communication and collegial exchange of ideas, centrally located functional support systems, and formal "front offices" and conference/presentation rooms.

The central location of the building's core led to a tri-part configuration. Design studios are located to the east, west, and south of the core, with the reception area, conference rooms, and principals' offices logically placed on the north, just off the elevator banks.

The main corridor, executed in black terrazzo, encircles the core and clearly defines the main circulation. A stair, consisting of two tubular trusses painted red, is visible from the reception area and indicates the presence of Murphy/Jahn on another floor.

墨菲/扬建筑师事务所

墨菲/扬建筑师事务所的办公室占用一个完整的楼层和另两个楼层的一部分空间。作为建筑师的办公室，其设计的出发点是在空间组织方面采用了古典原则。开放的设计工作室促进了自由交流和同事们交换意见，功能辅助系统集中设置，包括正式的前台办公室以及会议、演示用房。

该建筑物的核心筒位于其中央位置，从而产生了一个由三部分组成的构图形式。设计工作室位于核心筒体部分的东、西、南三面。将接待区、会议室和负责人的办公室按逻辑关系设置在其北侧，正好是在电梯的旁边。

采用黑色水磨石地面的主走廊环绕着建筑核心筒，并且清楚地限定了主要的交通流线。在接待区可以看到一部由两个涂红漆的管状桁架所构成的楼梯，它暗示着位于另一楼层上的墨菲/扬事务所的存在。

1

2

3

4

1	Circulation corridor	1	通行走道
2	Computer area	2	计算机区
3	Drafting area	3	草图工作区
4	Computer area	4	计算机区

Cupola

The cupola is located on the fortieth floor of the building occupied by Murphy/Jahn, Architects. Its original use is unknown, but it had been divided into a myriad of small spaces for use by a law firm.

Murphy/Jahn remodeled the space for use as a client presentation/conference area. This involved completely stripping the structure and recreating the entity of the dome. Four service modules were created around two axes of symmetry, and the centrally located smokestack which had to be retained. The floor was raised 3 feet to allow for better views through the windows which had unusually high parapets.

The cupola is reached by an elaborate, round, 1920s elevator reminiscent of the French cage type, which has been authentically restored.

穹顶屋

穹顶屋位于墨菲/扬建筑师事务所所在的大楼的第四十层上。无人知道该部分原来的用途，但它原来被划分成无数的小空间供一家律师事务所使用。

墨菲/扬建筑师事务所将该空间改造成了一个用于向客户展示汇报和会议的空间。该工作包括拆除分隔墙体和恢复穹顶的整体性。围绕两条对称的轴线形成了四个服务模块单元，并且不得不保留位于房间中央的巨大的排烟道。梯面被抬高了3英尺，以便能透过窗户更好地观赏城市的景色，而通常这种窗户设置了较高的窗下墙。

一部制作精美的圆形的20世纪20年代式的电梯可以将人们送到穹顶屋上，该电梯使人联想起法国的笼式电梯，长期以来，人们都忠实地维修着该电梯，使其保持了原貌。

5

5 穹顶屋：展示汇报区、分为三层的窗户和鸟笼式的电梯
6 穹顶屋：建筑物的中央大排烟道

5 Cupola: presentation area, 3 story windows and bird cage elevator
6 Cupola: central stack for building

6

Office of Murphy/Jahn, Architects 223

Biography

传记

Helmut Jahn FAIA

赫尔穆特·扬，FAIA

赫尔穆特·扬于1940年出生在德国的纽伦堡。1965年，他从慕尼黑工业大学毕业。一年后，他前往美国，到伊利诺伊工学院攻读硕士学位，得到了迈伦·戈德史密斯和法兹勒·卡恩的指导。

从1967年到1973年，他在C·F·墨菲联合事务所工作，成为吉恩·萨默尔斯的助手，于1973年成为执行副总裁和规划与设计部主任。1981年，他成为墨菲/扬事务所的一名主要负责人。第二年，他成为该事务所的总裁，并且在1983年还被任命为首席执行官。

时至今日，赫尔穆特·扬已经被公认为芝加哥最杰出的建筑师，他的作品极大地改变了芝加哥的城市面貌。按约翰·茹科夫斯基的评价，他的建筑作品对世界建筑产生了"令人震惊"的影响。他在国内和国际上不断扩大的影响使他获得了许多来自美国、加拿大、欧洲和远东的委托设计项目。他致力于设计建筑精品和改善城市环境。他设计的项目在设计创新、建筑的生命力和完整性方面获得了国际上的广泛认可。从无数关于他的作品的书籍中，读者们可以了解到他的作品在普通民众中和专业期刊杂志上所产生的兴奋感。墨菲/扬事务所的建筑赢得了无数的设计奖项，也在全球各地的建筑展览中得以展现。

他的设计作品既充满理性，又富于直觉；其设计试图赋予每栋建筑自身的哲学和智慧的基础，并形成一种开发其独特元素的机会，从而实现一种视觉上和交流式的叙述。其理性的部分用于使某个问题得以变为现实；而直觉方面则涉及到理论和智慧的层面——能够感受某个问题的内在结构和确定与空间、形式、光线、色彩和材料相关的各种设计元素的先后关系的潜意识上的能力，以及建筑通过建筑语言的符号和意义进行交流的方式。

他除了为事务所做各种工作外，还在伊利诺伊大学芝加哥校区任教，在哈佛大学担任建筑设计方面的埃利奥特·诺伊斯资助教授，在耶鲁大学担任建筑设计方面的达文波特资助客座教授，以及在伊利诺伊工学院担任论文教授。

Helmut Jahn was born in 1940 in Nürnberg, Germany. He graduated from the Technische Hochschule, Munich in 1965 and the following year moved to the United States to undertake graduate studies at the Illinois Institute of Technology, with Myron Goldsmith and Fazlur Kahn.

From 1967 to 1973 he worked with C.F. Murphy Associates as assistant to Gene Summers, being made Executive Vice President and Director of Planning and Design in 1973. In 1981 he became a Principal in Murphy/Jahn: the following year he became President of the practice, and in 1983 he was also made Chief Executive Officer.

Today, Helmut Jahn has been called Chicago's premiere architect who has dramatically changed the face of Chicago. His buildings have had a "staggering" influence on world architecture according to John Zukowsky. His growing national and international reputation has led to commissions across the United States, Canada, Europe and the Far East. He is committed to design excellence and the improvement of the urban environment. His projects have been recognized globally for design innovation, vitality and integrity. From the numerous publications on his work, one understands the excitement his work has generated in the public eye as well as professional journals and press. Murphy/Jahn's buildings have received numerous design awards and have been represented in architectural exhibitions around the world.

His design is both rational and intuitive; it attempts to give each building its own philosophical and intellectual base and establishes an opportunity to exploit its particular elements to achieve a visual and communicative statement. The rational part deals with the realities of a problem. The intuitive aspect deals with the theoretical, intellectual aspects—a subconscious ability to sense the intrinsic structure of a problem and establish priorities for the elements of design that deal with space, form, light, color and materials and the way architecture communicates through symbol and meaning of architectural language.

In addition to his work for the practice, he has taught at the University of Illinois Chicago Campus, was the Elliot Noyes Professor of Architectural Design at Harvard University and the Davenport Visiting Professor of Architectural Design at Yale University, and Thesis Professor at Illinois Institute of Technology.

Associates & Collaborators

共事人及合作者

Sam Scaccia
Executive Vice President, Director of Production
Sony Center, Berlin, Germany;
Principal Mutual Life Insurance Company, Des Moines, Iowa;
Charlemagne, Brussels, Belgium;
Victoria Berlin, Berlin, Germany;
Munich Airport Center, Munich, Germany;
Stralauer Platz 35, Berlin, Germany;
Kempinski Hotel, Munich, Germany;
21 Century Tower, Shanghai, China;
Ku-Damm 70, Berlin, Germany;
Pallas, Stuttgart, Germany;
United Airlines Terminal One Complex, Chicago, Illinois;
Messeturm, Frankfurt, Germany;
Northwestern Atrium Center, Chicago, Illinois;
425 Lexington Avenue, New York, New York;
One Liberty Place, Philadelphia, Pennsylvania;
120 North LaSalle, Chicago, Illinois;
Park Avenue Tower, New York, New York;
750 Lexington Avenue, New York, New York;
Cityspire, New York, New York;
Barnett Center, Jacksonville, Florida;
O'Hare Development Program, Chicago, Illinois;
Hyatt Regency Roissy, Paris, France;
Credit Lyonnais Bank Nederland, Rotterdam, The Netherlands;
300 East 85th Street, New York, New York

Philip Castillo
Senior Vice President, Principal Architect
21 Century Tower, Shanghai, China;
Sony Center Berlin, Germany;
Credit Lyonnais Bank Nederland, Rotterdam, The Netherlands;
Fort Canning Tower, Singapore;
Kongresshotel Messe, Frankfurt, Germany;
Tokyu-Shibuya Tower, Tokyo, Japan;
Kuala Lumpur City Center 2, Kuala Lumpur, Malaysia;
Lego Tower;
Wan-Chai Competition, Hong Kong;
Caltex Tower, Singapore;
Hitachi Tower, Singapore;
Celebration Center, Orlando, Florida;
Navy Pier, Chicago, Illinois;
MM21 Project, Yokohama, Japan;
Tokyo Teleport Town, Tokyo, Japan;
World's Tallest Building;
Dockland Square, London, England;
North Loop Block 37, Chicago, Illinois;
Citizens and Southern Bank, Atlanta, Georgia;
Wells & Lake Office Tower, Chicago, Illinois;
Niebelungenplatz, Frankfurt, Germany;
Oakbrook Terrace Tower, Oakbrook, Illinois;
362 West Street, Durban, South Africa;
Parktown Stands 85/87 and 102/103, Johannesburg, South Africa;
Merchandise Mart Bridge, Chicago, Illinois;
One South Wacker, Chicago, Illinois;
Madison Square Garden "Penn Gate", New York, New York

Brian J. O'Connor
Senior Vice President, Principal Architect
United Airlines Terminal One Complex, Chicago, Illinois;
O'Hare Development Program, Chicago, Illinois;
Flughafen Köln/Bonn, Cologne, Germany;
King Abdulaziz International Airport, Jeddah, Saudi Arabia;
Consolidated Terminal Project, JFK International Airport, Jamaica, New York;
American Airlines Terminal, JFK International Airport, Jamaica, New York;
Bangkok International Airport Competition, Bangkok, Thailand;
Okinawa Airport Competition, Okinawa, Japan;
Logan Airport Competition, Boston, Massachusetts;
Harold Washington Library Center Competition, Chicago, Illinois

Rainer Schildknecht
Senior Vice President, Principal Architect
Victoria Berlin, Berlin, Germany;
Stralauer Platz 35, Berlin, Germany;
Munich Airport Center, Munich, Germany;
Europa-Haus, Frankfurt, Germany;
Ku-Damm 119, Berlin, Germany;
Kempinski Hotel/Garage, Flughafen München 2, Munich, Germany;
Master Plan Neutral Zone, Flughafen München 2, Munich, Germany;
Ku-Damm 70, Berlin, Germany;
Munich Order Center, Munich, Germany;
Pallas, Stuttgart, Germany;
Mannheimer Lebensversicherung, Mannheim, Germany;
Messeturm, Frankfurt, Germany;
Messehalle, Frankfurt, Germany;
Northwestern Atrium Center, Chicago, Illinois;
Chicago Board of Trade Addition, Chicago, Illinois;
Area 2 Police Headquarters, Chicago, Illinois

Martin F. Wolf
Senior Vice President, Principal Architect
Flughafen Köln/Bonn, Cologne, Germany;
Principal Mutual Life Insurance Company, Des Moines, Iowa;
King Abdulaziz International Airport, Jeddah, Saudi Arabia;
Amtrak Station, Dearborn, Michigan;
Reichstagsgebäude Competition, Berlin, Germany;
Berlin 2000 Sports Complex, Berlin, Germany;
Bangkok International Airport Competition, Bangkok, Thailand;
American Airlines Terminal Complex, JFK International Airport, Jamaica, New York;
One America Plaza, San Diego, California;
Munich Airport Center, Munich, Germany;
First & Broadway, Los Angeles Civic Center, Los Angeles, California;
9th & Figueroa, Los Angeles, California;
Television City, New York, New York;
United Airlines Terminal One Complex, Chicago, Illinois;
Northwestern Atrium Center, Chicago, Illinois;
Bank of the Southwest Tower, Houston, Texas;
Naiman Sports Club, Chicago, Illinois;
Federal Life Insurance Company, Deerfield, Illinois;
O'Hare Development Program, Chicago, Illinois;
Monarch Beach Office Building, Dana Point, California;
Fountain Square West, Cincinnati, Ohio;
Chicago Board of Trade Addition, Chicago, Illinois;
FIS International Terminal Four, Chicago, Illinois;
Automated Guideway Transit Stations, Chicago, Illinois;
Security Bay Expansion, Chicago, Illinois;
Chicago Sports Complex, Chicago, Illinois

Associates & Collaborators Continued

Gordon Beckman
Vice President, Senior Project Architect
Jakarta Communications Tower,
Jakarta, Indonesia;
Victoria Berlin, Berlin, Germany;
Charlemagne, Brussels, Belgium;
Hamburg, Altona, Germany;
QCD, Mexico City, Mexico;
Salzufer, Berlin, Germany;
Alexanderplatz, Berlin, Germany;
Kaufhof Turm, Berlin, Germany;
MM21 Project, Yokohama, Japan;
Augustinerhof, Nuernberg, Germany;
CBC Tower, Toronto, Canada;
Place de Ville, Ottawa, Canada;
Wells and Lake Office Tower, Chicago, Illinois;
Hyatt Regency Roissy, Paris, France;
Amsterdam Master Plan, Amsterdam, The Netherlands;
Credit Lyonnais Bank Nederland, Rotterdam, The Netherlands;
TSP Competition Essen, Essen, Germany;
Kuala Lumpur City Center 2, Kuala Lumpur, Malaysia;
Barnett Center, Jacksonville, Florida;
Lake and Wells Office Tower, Chicago, Illinois;
750 Lexington Avenue, New York, New York;
Pacific Basin Tower, California;
Metro West, Naperville, Illinois

Steven S. Cook
Vice President, Senior Project Architect
Flughafen Köln/Bonn, Cologne, Germany;
Columbus Pier, St Pauli, Hamburg, Germany;
Munich Airport Center, Munich, Germany;
MUC 2 Walkways, Munich, Germany;
Munich Order Center, Munich, Germany;
Infrastruktur Flughafen München 2, Munich, Germany;
Kempinski Hotel and Parking Garage, Germany;
Pallas, Stuttgart, Germany;
Ku-Damm 70, Berlin, Germany;
1111 Brickell Avenue, Miami, Florida;
Mannheimer Lebensversicherung, Mannheim, Germany;
Mannheim-Ost: Tec Platz, Mannheim, Germany;
Ku-Damm 119, Berlin, Germany;
Messeturm, Frankfurt, Germany;
Madison Square Garden "Penn Gate", New York, New York;
First & Broadway, Los Angeles Civic Center, Los Angeles, California;
Park Avenue Tower, New York, New York

John S. Durbrow
Vice President, Senior Project Architect
Endless Towers, Singapore;
Jakarta Tower Kuningan Centre, Indonesia
Gehrlinghaus, Köln, Germany;
Mehringplatz, Berlin, Germany;
120 North LaSalle, Chicago, Illinois;
Europa-Haus, Frankfurt Germany;
Harold Washington Library Center Competition, Chicago, Illinois;
Victoria Berlin, Berlin, Germany;
Victoria Versicherung Competition, Düsseldorf, Germany;
Stuttgart/Berliner Platz, Stuttgart, Germany;
Hamburg Station Project, Hamburg, Germany;
Place de Ville III, Ottawa, Canada;
P&C/FWW Department Store, Berlin, Germany;
Hessische Landesbank Competition, Frankfurt, Germany;
Lufthansa Headquarters, Cologne, Germany;
Alliance Airport Terminal—Administration Building, Fort Worth, Texas;
Monmouth County Office Building, Monmouth, New Jersey;
Stralauer Platz 35, Berlin, Germany;
IHZ Tower, Düsseldorf, Germany;
Landesgirokasse Bank Center, Stuttgart, Germany;
Alexanderplatz, Berlin, Germany

Susan Froelich
Vice President, Senior Project Architect
21 Century Tower, Shanghai, China;
Sony Center Berlin, Germany;
Flughafen Köln/Bonn, Cologne, Germany;
King Abdulaziz International Airport, Jeddah, Saudi Arabia;
Tokyo Teleport Town, Tokyo, Japan;
Munich Airport Center, Munich, Germany;
Madison Square Garden "Penn Gate", New York, New York;
One America Plaza, San Diego California;
Ku-Damm 119, Berlin, Germany;
Messe Riem Urban Competition, Munich, Germany;
Monmouth County Office Building, Monmouth, New Jersey;
Northwestern Atrium Center West, Chicago, Illinois;
First & Broadway, Los Angeles Civic Center, Los Angeles, California;
Fort Canning Tower, Singapore;
North Loop Block 37, Chicago, Illinois;
Livingston Plaza, Brooklyn, New York;
Two Liberty Place, Philadelphia, Pennsylvania

Scott Pratt
Senior Vice President, Senior Project Architect
Sony Center Berlin, Germany;
Hyatt Regency Roissy, Paris, France;
One Liberty Place, Philadelphia, Pennsylvania;
Two Liberty Place, Philadelphia, Pennsylvania;
Wilshire/Westwood, Los Angeles, California;
9th & Figueroa, Los Angeles, California;
1000 West Sixth Street, Los Angeles, California;
Park Place, Los Angeles, California;
West Loop Office Project, Chicago, Illinois;
Houston Tower, Houston, Texas;
Northbrook Centre, Northbrook, Illinois;
West Loop Pedway, Chicago, Illinois;
701 Fourth Avenue South, Minneapolis, Minnesota;
One Hennepin Center, Minneapolis, Minnesota;
The Esplanade, Houston, Texas;
Hyatt Regency Frankfurt am Main, Frankfurt, Germany;
University of Illinois Agricultural Engineering Sciences Building, Urbana-Champaign, Illinois;
First Source Center, South Bend, Indiana;
W.W. Grainger Corporate Headquarters, Skokie, Illinois;
Oakbrook Post Office, Oakbrook, Illinois

Dennis Recek (1968–1990)
Vice President, Senior Project Architect
120 North LaSalle, Chicago, Illinois;
One America Plaza, San Diego, California;
Wilshire/Westwood, Los Angeles, California;
1000 West Sixth Street, Los Angeles, California;

Harold Washington Library Center
Competition, Chicago, Illinois;
Metro West, Naperville, Illinois

Dieter Zabel
Vice President, Senior Project Architect
Mehringplatz, Berlin, Germany;
Stralauer Platz 35, Berlin, Germany;
Sony Center Berlin, Germany;
Grossmarkthalle, Frankfurt, Germany;
Reichstagsgebäude Competition, Berlin, Germany;
Berlin 2000 Sports Complex, Berlin, Germany;
Regierungsviertel Spreebogen Competition, Berlin, Germany;
Urban Competition Alexanderplatz, Berlin, Germany;
Europa-Haus, Frankfurt, Germany;
Ku-Damm 119, Berlin, Germany;
Kongresshotel Messe, Frankfurt, Germany;
Messe Riem Urban Competition, Munich, Germany;
Munich Airport Center Infrastruktur, Munich, Germany;
P&C/FWW Department Store, Berlin, Germany;
Victoria Versicherung Competition, Düsseldorf, Germany;
Helaba, Frankfurt, Germany;
Messeturm, Frankfurt, Germany;
Messehalle, Frankfurt, Germany;
Mannheimer Lebensversicherung, Mannheim, Germany

Edward P. Wilkas
Vice President, Director of Production
Principal Mutual Life Insurance Company, Des Moines, Iowa;
120 North LaSalle, Chicago, Illinois;
Hyatt Regency Roissy, Paris, France;
North Loop Block 37, Chicago, Illinois;
GSA Office Building Competition, Chicago, Illinois;
West Loop Pedway, Chicago, Illinois;
One Liberty Place, Philadelphia, Pennsylvania;
State of Illinois Center, Chicago, Illinois;
Randolph Street Pedway, Chicago, Illinois;
Lake/Clark Transfer Station, Chicago, Illinois

Mark Joshua Frisch
Director of Production
Victoria Berlin, Berlin, Germany;
Europa-Haus, Frankfurt, Germany;
Ku-Damm 119, Berlin, Germany;
Munich Order Center, Munich, Germany;
Hyatt Regency Roissy, Paris, France;
Messeturm, Frankfurt, Germany;
Messehalle, Frankfurt, Germany;
Two Liberty Place, Philadelphia Pennsylvania;
One Liberty Place Competition, Philadelphia, Pennsylvania;
Cityspire, New York, New York;
Park Avenue Tower, New York, New York;
O'Hare Development Program, Chicago, Illinois

Sanford E. Gorshow
Director of Production
Stralauer Platz 35, Berlin, Germany;
Principal Mutual Life Insurance Company, Des Moines, Iowa;
Europa-Haus, Frankfurt, Germany;
Munich Order Center, Munich, Germany;
Monarch Beach Office Building, Dana Point, California;
Pallas, Stuttgart, Germany;
Ku-Damm 70, Berlin, Germany;
Mannheimer Lebensversicherung, Mannheim, Germany;
United Airlines Terminal One Complex, Chicago, Illinois;
Shand Morahan, Evanston, Illinois;
Plaza East, Milwaukee, Wisconsin

Stephen Kern
Director of Production
21 Century Tower, Shanghai, China;
Sony Center Berlin, Berlin, Germany;
Credit Lyonnais Bank Nederland, Rotterdam, The Netherlands;
Hyatt Regency Roissy, Paris, France;
Messeturm, Frankfurt, Germany;
Messehalle, Frankfurt, Germany;
Merchandise Mart Bridge, Chicago, Illinois;
One Liberty Place, Philadelphia, Pennsylvania;
425 Lexington Avenue, New York, New York;
362 West Street, Durban, South Africa;
Northwestern Atrium Center, Chicago, Illinois

Steven Michael Nilles
Director of Production
Sony Center, Berlin, Germany;
Munich Airport Center, Munich, Germany;
Kempinski Hotel, Munich, Germany;
MUC 2 Tiefgarage, Munich, Germany;
Livingston Plaza, Brooklyn, New York;
Barnett Center, Jacksonville, Florida;
Northern Trust Pedestrian Bridge, Chicago, Illinois;
Messehalle, Frankfurt, Germany;
Wilshire/Westwood, Los Angeles, California;
United Airlines Terminal One Complex, Chicago, Illinois

Thomas M. Chambers
Project Architect
Flughafen Köln/Bonn, Cologne, Germany;
Midway Terminal Improvement Program, Chicago, Illinois;
Bangkok International Airport Competition, Bangkok, Thailand;
United Airlines Terminal One Complex, Chicago, Illinois;
King Abdulaziz International Airport, Jeddah, Saudi Arabia;
American Airlines Terminal Complex, JFK International Airport, Jamaica, New York;
O'Hare Development Program, Chicago, Illinois;
Consolidated Terminal Project, JFK International Airport, Jamaica, New York;
Mannheimer Lebensversicherung, Mannheim, Germany;
Monarch Beach Office Building, Dana Point, California;
Harold Washington Library Center Competition, Chicago, Illinois;
Park Place Competition, Los Angeles, California;
Italian Sports Complex, Rome, Italy

Consulting Architects
Charles Bostick
Fritz Ludwig
Lothar Pascher
Raimund Schock

Former Senior Personnel
Daniel Dolan
James Goettsch
Robert Goldberg
James M. Stevenson

Chronological List of Buildings & Projects
(By year of project commencement)

*Indicates work featured in this book
(see Selected and Current Works).

Mat-Buildings

建筑及项目年表

H. Roe Bartle Exhibition Hall
Kansas City, Missouri
Design 1972/Completion 1976
C.F. Murphy Associates
In joint venture with:
    Seligson Associates
    Hormer and Blessing
    Howard Needles Tammen Bergendoff

*Kemper Arena
Kansas City, Missouri
Design 1973/Completion 1974
American Royal Arena Corporation

John Marshall Courts Building
Richmond, Virginia
Design 1973/Completion 1976
Richmond Court Facility

Fourth District Courts Building
Maywood, Illinois
Design 1974/Completion 1976
Cook County Building Commission

*Auraria Library
Denver, Colorado
Design 1974/Completion 1976
Auraria Higher Education District

*Michigan City Public Library
Michigan City, Indiana
Design 1974/Completion 1977
Michigan City Public Library System

*Saint Mary's Athletic Facility
South Bend, Indiana
Design 1976/Completion 1977
Saint Mary's College

*Rust-Oleum Corporation Headquarters
Vernon Hills, Illinois
Design 1976/Completion 1979
Rust-Oleum Corporation

Commonwealth Edison District Headquarters
Bolingbrook, Illinois
Design 1976/Completion 1979
Commonwealth Edison

*De La Garza Career Center
East Chicago, Indiana
Design 1976/Completion 1981
City of East Chicago

*Minnesota Government & History Center
St Paul, Minnesota
Design 1976
State of Minnesota

Mat-Buildings continued

*****Abu Dhabi Conference Center City**
United Emirates
Design 1976
United Arab Emirates

La Lumiere Gymnasium
La Porte, Indiana
Design 1977/Completion 1978
La Lumiere School

Oakbrook Post Office
Oakbrook, Illinois
Design 1978/Completion 1981
United States Post Office

*****Area 2 Police Headquarters**
Chicago, Illinois
Design 1978/Completion 1981
City of Chicago

*****ANL/DOE Program Support Facility**
Argonne, Illinois
Design 1978/Completion 1981
US Department of Energy

**University of Illinois
Ag. Eng. Science Building**
Urbana-Champaign, Illinois
Design 1978/Completion 1984
State of Illinois, Capital Development Board

COD/Learning Resources Center
Glen Ellyn, Illinois
Design 1979/Completion 1983
State of Illinois, Capital Development Board

*Wisconsin Residence
Eagle River, Wisconsin
Design 1980/Completion 1982
Private residence

Codex Corporate Headquarters
Canton, Massachusetts
Design 1982
Codex Corporation

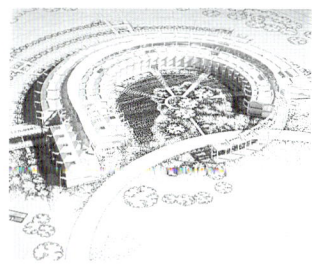

Exposition and Office Complex
Chicago, Illinois
Design 1982

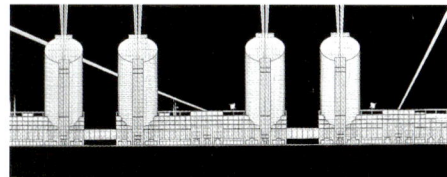

Parktown Stands
Johannesburg, South Africa
Design 1984/Completion 1986
Anglo American Properties
Associate Architect:
    John Kemp & Associates

*San Diego Convention Center
San Diego, California
Design 1984
San Diego Convention Bureau
In joint venture with:
    Martinez/Wong Associates, Inc.

Mat-Buildings continued

Messe Hall 1
Design 1985/Completion 1988
Messe Hall Messe Frankfurt GmbH

Palace Prince Khaled
Saudi Arabia
Design 1987
Private Residence

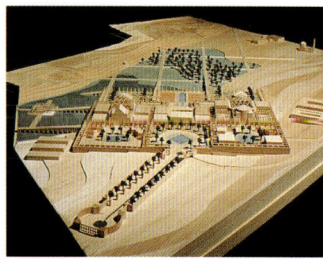

*Munich Order Center
Munich, Germany
Design 1989/Completion 1993
Archimedes Gewerbe - und Buero Centrum
    GmbH & Co.

Monarch Beach Office Building
Dana Point, California
Design 1989
Stein Brief Partnership

*Navy Pier
Chicago, Illinois
Design 1991
Metropolitan Pier and Exhibition Authority

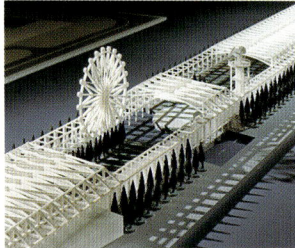

*Celebration Center
Florida
Design 1991
Disney Development Co.

Messe Riem Urban Competition
Munich, Germany
Design 1992

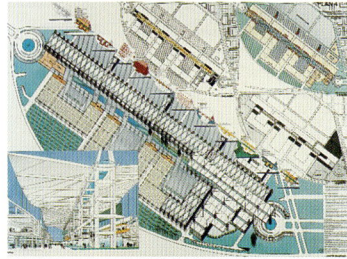

Tower Buildings

*Xerox Centre
Chicago, Illinois
Design 1978/Completion 1980
Romanek-Golub

*Chicago Board of Trade Addition
Chicago, Illinois
Design 1978/Completion 1982
Chicago Board of Trade
In joint venture with:
    Shaw and Associates
    Swanke Hayden Connell Architects

*One South Wacker
Chicago, Illinois
Design 1979/Completion 1982
Metropolitan Structures/
    Harvey Walken & Co.

Plaza East
Milwaukee, Wisconsin
Design 1980/Completion 1982 (1)
Design 1981/Completion 1984 (2)
JMB Reality

*11 Diagonal Street
Johannesburg, South Africa
Design 1981/Completion 1984
Anglo American Properties
Associate Architect:
    Louis Karol Architects

701 Fourth Avenue South
Minneapolis, Minnesota
Design 1982/Completion 1984
Turner Development Corporation

Tower Buildings continued

*Park Avenue Tower
New York, New York
Design 1982/Completion 1986
Park Tower Realty

*362 West Street
Durban, South Africa
Design 1983/Completion 1986
Anglo-American Properties
Associate Architect:
    Stauch Vorster + Partners

*Metro West
Naperville, Illinois
Design 1984/Completion 1986
Westminister Corporation

*Northwestern Atrium Center
Chicago, Illinois
Design 1984/Completion 1987
Tishman Midwest Management Corp

*One Liberty Place
Philadelphia, Pennsylvania
Design 1984/Completion 1987
Rouse Associates

*Wilshire/Westwood
Los Angeles, California
Design 1984/Completion 1988
Developer: Platt Corporation
Owner: The Prospect Co.

*Cityspire
New York, New York
Design 1984/Completion 1989
West 56th Street Associates

*750 Lexington Avenue
New York, New York
Design 1984/Completion 1989
Cohen Brothers

*Oakbrook Terrace Tower
Oakbrook Terrace, Illinois
Design 1985/Completion 1987
Miglin Beitler Incorporated

300 East 85th Street
New York, New York
Design 1985/Completion 1988
East 85th Street Associates

*425 Lexington Avenue
New York, New York
Design 1985/Completion 1989
Olympia & York Equity Corporation

*Messe Tower/Messe Hall
Frankfurt, Germany
Design 1985/Completion 1991
TishmanSpeyer Properties of Germany, L.P.

Tower Buildings continued

*****Bank of Southwest Tower**
Houston, Texas
Design 1985
Century Development Corporation
Associate Architect:
    Lloyd Jones Brewer Associates

Pacific Basin Tower
Design 1985

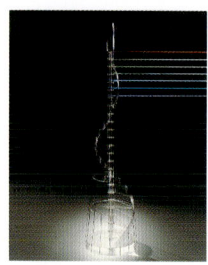

Television City
New York, New York
Design 1985
Trump Organization

9th & Figueroa
Los Angeles, California
Design 1985
RCI

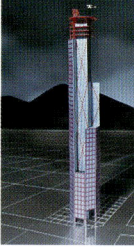

*****Columbus Circle**
New York, New York
Design 1985
Trump Organization

*****South Ferry Plaza**
New York, New York
Design 1986
Olympia & York Equity Corporation

Stuttgart/Berliner Platz
Stuttgart, Germany
Design 1986
OPUS Entwicklungsgesellschaft

1000 West Sixth Street
Los Angeles, California
Design 1987
Platt Corporation

*Barnett Center
Jacksonville, Florida
Design 1988/Completion 1990
Developer: Paragon Group, Inc.
Owner: Barnett Bank

*120 North LaSalle
Chicago, Illinois
Design 1988/Completion 1991
Ahmanson Commercial
    Development Corporation

Two Liberty Place
Philadelphia, Pennsylvania
Design 1988/Completion 1991
Rouse Associates
Office Design Architect: Murphy/Jahn
Architect: Zeidler Roberts Partnership

One America Plaza
San Diego, California
Design 1988/Completion 1993
Shimizu Corporation

Chronological List of Buildings & Projects

Tower Buildings continued

*Hitachi Tower
Singapore
Design 1988/Completion 1993
Savu Properties Pte.
Associate Architect: Architects 61

*Caltex House
Singapore
Design 1989/Completion 1993
Savu Properties Pte.
Associate Architect: Architects 61

Credit Lyonnais Bank Nederland
Rotterdam, The Netherlands
Design 1988/Completion 1995
Credit Lyonnais Bank Nederland
Associate Architect: Inbo Architektenburo

*The North Loop Block 37
Chicago, Illinois
Design 1988
FJV Venture

*1111 Brickell Avenue
Miami, Florida
Design 1988
1111 Brickell Associates

*Project J (1, 2, 3)
Chicago, Illinois
Design 1989
Howard Ecker & Co.

Wan-Chai Competition
Hong Kong
Design 1989
Sun Hung Kai Properties

Citizens & Southern Bank
Atlanta, Georgia
Design 1989
Citizens & Southern Bank

Canadian Broadcast Tower
Toronto, Canada
Design 1989

***Tokyo Teleport Town**
Tokyo, Japan
Design 1990
Associate Architect: Kajima Corporation

World's Tallest Building
Design 1990

Dockland Square
London, England
Design 1990
Olympia & York

Tower Buildings continued

*Yokohama Waterfront MM21 Project
Yokohama, Japan
Design 1990
Mitsui & Co.
Associate Architect:
    A.A.S. Associates Intíl. Co.

*Kuala Lumpur City Center 1
Kuala Lumpur, Malaysia
Design 1991
Seri Kuda Sdn. Bhd.

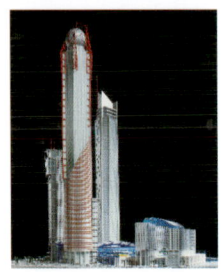

*FAA O'Hare Control Tower
Chicago, Illinois
Design 1991
Chicago O'Hare International Airport
Dept. of Aviation, City of Chicago

*Fountain Square West
Cincinnati, Ohio
Design 1991
The Galbreath Company

*IHZ Tower
Düsseldorf, Germany
Competition 1992
Premier GmbH Bouygues

*Fort Canning Tower
Singapore
Design 1992
Singapore Telecom
Associate Architect: Architects 61

*Tokyu-Shibuya Tower
Tokyo, Japan
Design 1992
Tokyu Corporation
Associate Architects:
    Kume Sekkei Co, Ltd
    Media Five

*21 Century Tower
Shanghai, China
Design 1993
Completion 1997
China Everbright International Trust
    and Investment Corp.
Associate Architect:
    East China Architectural Design Institute

*Jakarta Tower Kuningan Centre
Jakarta, Indonesia
Design 1994
Pacific Metrorealty

*Endless Towers
Singapore
Design 1995
Pidemco Land Limited
Associate Architect: Architects 61

*Jakarta Communications Tower
Jakarta, Indonesia
Design 1995
PT Kuningan Persada

Urban Block Buildings

获奖情况及展览

*First Source Center
South Bend, Indiana
Design 1978/Completion 1982
First Source Bank
Rahn Properties
City of South Bend

*State of Illinois Center
Chicago, Illinois
Design 1979/Completion 1985
State of Illinois, Capital Development Board
In joint venture with:
    Lester B. Knight & Associates

*Shand Morahan Corporate Headquarters
Evanston, Illinois
Design 1982/Completion 1984
Walken Souyoul Interests

Hague Town Hall and Library
Amsterdam, The Netherlands
1986

*First & Broadway Los Angeles Civic Center
Los Angeles, California
Design 1987
RCI

Livingston Plaza
Brooklyn, New York
Design 1988/Completion 1991
Developer: Cohen Brothers Realty
Client: NYCT

***Mannheimer Lebensversicherung Headquarters**
Mannheim, Germany
Design 1988/Completion 1992
Mannheimer Lebensversicherung AG
Mannheimer Versicherung AG

***Ku-Damm 70**
Berlin, Germany
Design 1988/Completion 1994
Euwo Unternehmensgruppe

***Ku-Damm 119**
Berlin, Germany
Design 1988/Completion 1995
Athena Grundstuecks AG
Vebau GmbH Frankfurt/Berlin

***Harold Washington Library Center**
Competition
Chicago, Illinois
Design 1988
City of Chicago Library System,
   Tishman Midwest

***Victoria Düsseldorf Competition**
Düsseldorf, Germany
Design 1988
Victoria Versicherungsgesellschaft

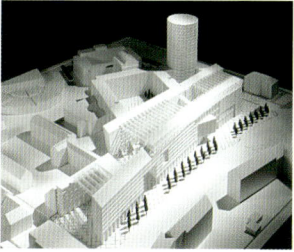

***Pallas Office Building**
Stuttgart, Germany
Design 1989/Completion 1994
IGEPA - Industrie - und
   Gewerbepark Bautraeger

Chronological List of Buildings & Projects   243

Urban Block Buildings continued

*Augustinerhof-Nuernberg
Nuernberg, Germany
Design 1990/Competion 1997
Abousaidy

*Lufthansa Corporate Headquarters
Cologne, Germany
Design 1990
Lufthansa German

*Victoria Versicherungsgesellschaft
Berlin, Germany
Design 1992/Completion 1997
Victoria Versicherungsgesellschaft

*Europa-Haus
Frankfurt, Germany
Design 1992
SkanInvest

Landesgirokasse Bank Center Competition
Stuttgart, Germany
Design 1992
Landesgirokasse Bank

Communication Center FFM
Frankfurt, Germany
Design 1992

***Principal Mutual Life
Insurance Company**
Des Moines, Iowa
Design 1993/Completion 1995
Principal Mutual Life Insurance Company

*Stralauer Platz 35
Berlin, Germany
Design 1993/Completion 1997
OPUS Entwicklungsgesellschaft

Columbus Pier
Hamburg, Germany
Design 1993/Completion 1997
Garbe Projekt-Entwicklung

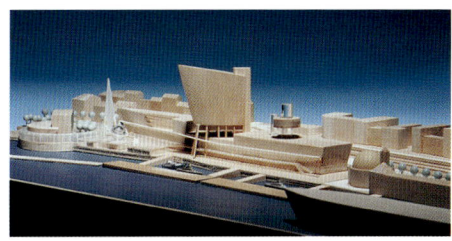

*Sony Center, Berlin
Berlin, Germany
Design 1993/Completion 1998
Sony Corporation

Reichstagsgebaude Competition
Berlin, Germany
Design 1993

*Charlemagne
Brussels, Belgium
Design 1994
Etudes et Investissements
    Immobiliers, S.A./N.V.
Associate Architect: ARC, S.A. Architectes

Transportation Buildings

*O'Hare Rapid Transit Station
Chicago, Illinois
Design 1979/Completion 1984
Chicago Building Commission

*United Airlines Terminal One Complex
Chicago-O'Hare International Airport
Design 1983/Completion 1987
United Airlines

Merchandise Mart Bridge
Chicago, Illinois
Design 1987/Completion 1988
Merchandise Mart Properties

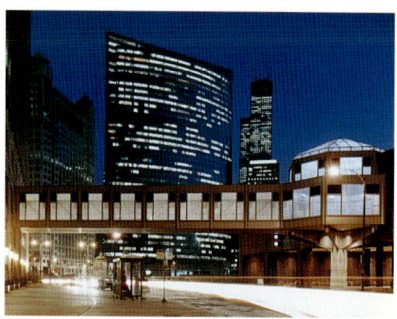

Alliance Airport Terminal, Administration Building
Fort Worth, Texas
Design 1988
Alliance Airport

West Loop Pedway
Chicago, Illinois
Design 1989
City of Chicago

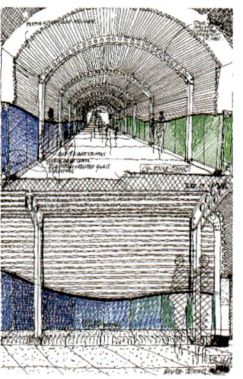

*One America Plaza/Trolley Station
San Diego, California
Design 1988/Completion 1991
Shimizu Corporation

*Hyatt Regency Roissy
Paris, France
Design 1988/Completion 1992
Hyatt International Corporation
Toa Investment Pte., Ltd

*JFK Consolidated Terminal
Schematic Design
Jamaica, New York
Design 1988
American Airlines/Port Authority of NY/NJ

*Hotel Kempinski/Munich Airport Center
/Walkways
Munich, Germany
Hotel Design 1989/Completion 1994
MAC Design 1990/Completion 1997
Walkways Design 1991/Completion 1992
Hotel/Walkways: Flughafen Munchen GmbH
MAC: Flughafen Munchen GmbH
   MFG Beta KG
   ALBA GmbH

Transportation Buildings continued

*King Abdulaziz International Airport
Competition, Conceptual Design,
Schemes 1 & 2
Jeddah, Saudi, Arabia
Design 1991/1993
International Airport Projects
Ministry of Defense and Aviation

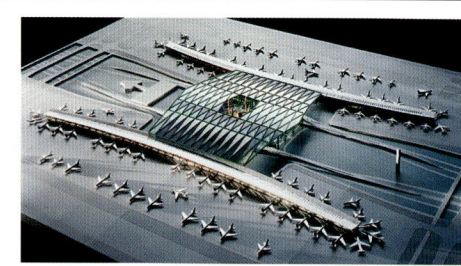

*Flughafen Köln/Bonn Terminal 1
Cologne, Germany
Design 1992/Completion 2000
Flughafen Köln/Bonn GmbH

*Dearborn Station
Dearborn, Michigan
Design 1994
Union Station Company

*Bangkok International Airport
Bangkok, Thailand
Design 1994/Completion 2001
Airports Authority of Thailand

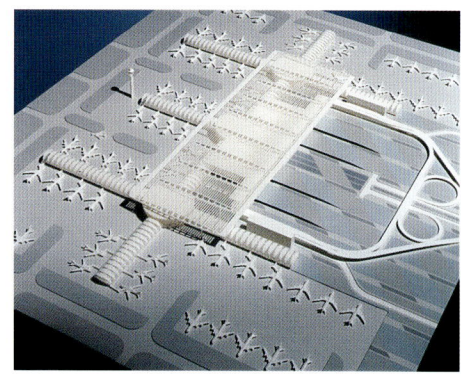

Awards & Exhibitions

Awards

AIA Chicago Chapter Award
Hyatt Regency Roissy
1994

AIA Chicago Chapter Award
Munich Order Center
1994

Outstanding Achievement/Architect
Illinois Academy Of Fine Arts
1993

New Chicago Architecture Award
120 North LaSalle
1993

New Chicago Architecture Award
Sony/Berlin
1993

CIDB Best Buildable Design Awards
Hitachi Tower
1993

Bund Deutscher Architektan
Mannheimer Lebensversicherung
1993

IIT Award
For Outstanding Contribution to the Built Environment
1992

Chicago Architecture Award
American Institute Of Architects, Illinois Chapter/Merchandise Mart
1992

AIA San Diego Chapter Award
One America Plaza Trolley Station
1992

AIA Chicago Chapter Award
120 North LaSalle
1992

AIA Chicago Chapter Award
Mannheimer Lebensversicherung
1992

Ten Most Influential Living American Architects
American Institute of Architects
Helmut Jahn
1991

Ten Best Works of American Architecture Completed Since 1980
American Institute of Architects
State Of Illinois Center
1991

Ten Best Works of American Architecture Completed Since 1980
American Institute of Architects
United Airlines Terminal
1991

Development of the Year
Chicago Sun Times Real Estate
120 North LaSalle
1991

Dean of Architecture Award
Chicago Design Awards
1991

AIA Chicago Chapter Award
Messe Halle
1991

AIA Chicago Chapter Award
Wilshire/Westwood
1991

"Top 100 Architects"
Architectural Digest
1991

"The Top Ten" Chicago's Ten Most Important Post World War II Works of Architecture
State Of Illinois Center
Paul Gapp
1991

Quaternario International Award for Innovative Technology in Architecture
United Airlines Terminal
1990

Domino's 30 Award
1990

AISC Award
United Airlines Terminal
1990

AISC Award
One Liberty Place
1990

Divine Detail Honor Award
American Institute of Architects, Chicago Chapter
United Airlines Terminal
1990

AIA Chicago Chapter Award
One Liberty Place
1990

Domino's 30 Award
1989

American Consulting Engineers Council Award
United Airlines Terminal
1989

R.S. Reynolds Memorial Award
Annual Award for Distinguished Architecture Using Aluminum
United Airlines Terminal
1988

NEA Presidential Design Award
O'Hare Rapid Transit Station
1988

Chevalier Dans L'Ordre des Arts et des Lettres
Ministere de la Culture et de la Communication
Paris, France
Helmut Jahn
1988

Annual Design Review "Best of Category"
Industrial Design Magazine
United Airlines Terminal
1988

Twenty-Five Year Award
American Institute of Architects, Chicago Chapter
O'Hare International Airport
1988

AIA Chicago Chapter Award
United Airlines Terminal
1988

Structural Engineering Association of Illinois Award
United Airlines Terminal
1987

Annual Award
Concrete Industry Board of New York
Cityspire
1987

AIA National Honor Award
O'Hare Rapid Transit Station
1987

AIA National Honor Award
United Airlines Terminal
1987

AIA Chicago Chapter Award
Metro West
1987

Structural Engineering Association of Illinois Award
Northwestern Atrium Center
1986

AIA New York State Award
701 Fourth Avenue South
1986

AIA Chicago Chapter Award
State Of Illinois Center
1986

Structural Engineering Association of Illinois Award
Addition to the Chicago Board of Trade
1985

Structural Engineering Association of Illinois Award
State Of Illinois Center
1985

Progressive Architecture Award
Chicago Central Area Plan
1985

Awards & Exhibitions Continued

Distinguished Architect
Helmut Jahn
City of Milwaukee Art Commission's 1985
Annual Awards
Plaza East Office Center
1985

AIA Chicago Chapter Award
Addition to the Chicago Board of Trade
1984

AISC Award
Addition to the Chicago Board of Trade
1983

AIA Chicago Chapter Award
Area 2 Police Headquarters
1983

Second Annual Award for Distinguished Architecture
Reliance Development Group Inc.
Addition to the Chicago Board of Trade
1982

Arnold W. Brunner Memorial Prize in Architecture
Helmut Jahn
1982

AIA Chicago Chapter Award
Argonne Program Support Facility
1982

ASHRAE Energy Award
De La Garza Career Center
1981

ASHRAE Energy Award
Commonwealth Edison District Headquarters
1981

ASHRAE Energy Award
State Of Illinois Center
1981

AIA Chicago Chapter Award
De La Garza Career Center
1981

AIA Chicago Chapter Award
Xerox Centre
1980

Young Professional Award
Building Design & Construction Magazine
Rust-Oleum Headquarters
1979

Owens-Corning Fiberglas Energy Conservation Award
Argonne Program Support Facility
1979

AISC Award
Michigan City Library
1979

AISC Award
Rust-Oleum Headquarters
1979

AIA National Honor Award
St Mary's Athletic Facility
1979

AIA Chicago Chapter Award
Rust-Oleum Headquarters
1979

Progressive Architecture Design Citation
Minnesota II
1978

AISC Award
St Mary's Athletic Facility
1978

AIA Chicago Chapter Award
Minnesota II
1978

AIA Chicago Chapter Award
Kansas Convention Center
1978

AIA Chicago Chapter Award
La Lumiere Gymnasium
1978

AIA Chicago Chapter Award
Oak Brook Post Office
1978

AIA Chicago Chapter Award
University of Illinois Agricultural Engineering Science Building
1978

AIA American Library Association First Honor Award
Michigan City Library
1978

Winner in National Architecture Competition
Minnesota II
1977

Progressive Architecture Design Citation
Xerox Centre
1977

AIA Illinois Council Honor Award
Michigan City Library
1977

AIA Illinois Council Honor Award
St Mary's Athletic Facility
1977

AIA Chicago Chapter Award
Michigan City Library
1977

AIA Chicago Chapter Award
St Mary's Athletic Facility
1977

Progressive Architecture Design Citation
Abu Dhabi Conference City
1976

Prize in International Competition
Abu Dhabi Conference City
1976

AIA Chicago Chapter Award
Auraria Library
1976

Bartelt Award
Kemper Arena
1975

AISC Award
Kemper Arena
1975

AIA National Honor Award
Kemper Arena
1975

AIA Chicago Chapter Award
Kemper Arena
1975

Exhibitions

Building for Air Transportation
Art Institute of Chicago
1995

New Chicago Architecture,
The Chicago Athenaeum
November 1994

Hyatt Amsterdam
ARCAM Galerie, Amsterdam,
The Netherlands
January 1994

New Chicago Architecture Exhibition
Europa-Haus, Fort Canning Tower, Hyatt Regency Roissy, Messeturm/Messe Halle, 120 North LaSalle, Sony/Berlin
March 1993

Chicago Architecture and Design: 1923–1993
Art Institute of Chicago
May 1993

The Art of Design
Milwaukee, Wisconsin
June 1993

City-Projects
Berlin Pavilion, Berlin, Germany
September 1993

Buenos Aires Biennial of Architecture '93
Recent Work at the Museum of Fine Arts
September 1993

Celebrating 75 Years of Chicago Architecture
The Arts Club of Chicago
January 1992

From Mars to Main Street: America Designs
Washington, DC
November 1992 to December 1993

American Skyscrapers
Helsinki, Finland
March 1993

New Chicago Architecture
The Chicago Athenaeum
January 1991

Helmut Jahn
YKK—Cupples Design Forum, Tokyo, Japan
November 1991

Chicago Architecture
Gulbenkian Foundation
Lisbon, Portugal
1990

The Socially Responsible Environment: US/USSR 1980–1990
Architects Designers Planners for Social Responsibility, USA, and USSR Union of Architects
New York, New York
May 1990

Celebrate Chicago Architecture
Metropolitan Press
Union Station in Washington, DC
June 1990

Terminus to Interchange
The Building Centre Trust, London, England
September 1990

The Art of Design
One Liberty Place, Bank of the Southwest Tower
STA Gallery, Chicago, Illinois
January 1989

Architecture in New York
Columbus Circle, Times Square, Terminalgebäude für AA, North Western JFK
Deutsches Architekturmuseum, Germany
March 1989

Original Unpublished Drawings and Models by Chicago Architectural Club Members
Ruth Volid Gallery, Chicago, Illinois
June 1989

Design USA—A US–USSR Cultural Exchange Exhibition
United States Information Agency, USSR
September 1989

Competitions in Architecture
The Architectural League of New York
New York, New York
September 1989

Tall Buildings
Brisbane, Australia
West Perth, Australia
October 1989

New Chicago Architecture
Metropolitan Press, Chicago, Illinois
November 1989

The Experimental Tradition
The Architectural League, New York, New York
May 1988

New Chicago Skyscrapers
ArchiCenter Gallery, Chicago, Illinois
July 1988

Experimental Skytowers
Rizzoli International Bookstore, Chicago, Illinois
August 1988

The Cities of the World and the Future of the Metropolis
Triennale di Milano, Italy
September 1988

The Postmodern Explained to Children
Bonne Fanten Museum, Maastricht, The Netherlands
November 1988

What Could Have Been: Unbuilt Architecture of the '80s
Dallas Market Center, Dallas, Texas
January 1987

Chicago Architecture Foundation,
Chicago, Illinois
December 1987

GA International '87
GA Gallery, Tokyo, Japan
May 1987

Helmut Jahn
Architekturgalerie, Munich, Germany
June 1987

New Chicago Skyscrapers
Merchandise Mart, Chicago, Illinois
July 1987

Helmut Jahn
Paris Art Center, Paris, France
September 1987

Members' Work
The Chicago Architectural Club
Betsy Rosenfield Gallery, Chicago, Illinois
January 1986

Modernism Redux: Critical Alternatives
Grey Art Gallery and Study Center, New York University, New York, New York
February 1986

Vision der Moderne
Deutsches Architekturmuseum, Frankfurt, Germany
June 1986

The National Museum of Modern Art, Tokyo
October 1986

Helmut Jahn
Gallery MA, Tokyo, Japan
October 1986

New Chicago Skyscrapers
Gallery of Design of the Merchandise Mart, Chicago, Illinois
October 1986

Exhibition on Advanced Structures
Syracuse University, Syracuse, New York
January 1985

Vu de l'Interieur ou la Raison de l'Architecture
(Looking from the Inside or the Reason of Architecture)
Paris Biennale, Paris, France
March 1985

150 Years of Chicago Architecture
Museum of Science and Industry, Chicago, Illinois
October 1985

Helmut Jahn
Ballenford Architectural Books and Gallery
Toronto, Ontario, Canada
February 1984

Chicago and New York, More than a Century of Architectural Interaction
The Art Institute of Chicago; The AIA Foundation, The Octagon, Washington, D.C.; Farish Gallery, Rice University, Houston, Texas; The New York Historical Society, New York, New York
March 1984

Opening, Permanent Architectural Collection
Deutsches Architekturmuseum, Frankfurt, Germany
June 1984

Art and... Architecture/Design
Moosart Gallery, Miami, Florida
November 1984

Members' Work
The Chicago Architectural Club
The Art Institute of Chicago, Chicago, Illinois
November 1984

The State of Illinois Center
University of Illinois, Champaign-Urbana, Illinois
December 1984

The Architect's Vision From Sketch to Final Drawing
The Chicago Historical Society, Chicago, Illinois
February 1983

Current Projects
Thomas Beeby, Lawrence Booth, Helmut Jahn, Krueck and Olson, Stanley Tigerman
Young-Hoffman Gallery Chicago, Illinois
February 1983

Awards & Exhibitions Continued

Ornamentalism: The New Decorativeness in Architecture and Design
The Hudson River Museum, New York, New York; Archer M. Huntington Art Gallery, University of Texas, Austin, Texas
March 1983

New Chicago Architecture 1983
The Art Institute of Chicago, Chicago, Illinois
May 1983

Minneapolis Profile 1983
Walker Art Center, Minneapolis, Minnesota
Participant in symposium
May 1983

Tall Buildings
The Southern Chapter, Alberta Association of Architects, Calgary, Alberta, Canada
June 1983

Design USA
Castello Sforzasco in the Sala Viscontea, Milano, Italy
September 1983

Surface and Ornament
Puck Building, New York, New York
October 1983

Contemporary Landscape—from the Horizon of Postmodern Design
The National Museums of Modern Art, Kyoto and Tokyo, Japan
September/December 1985

Competitions Won and Lost
San Francisco AIA Headquarters Gallery, San Francisco, California
October 1983

150 Years of Chicago Architecture
Paris Art Center, Paris, France
October 1983

1992 Chicago World's Fair Design Conference Drawings, New York, Los Angeles, Chicago
The University of Illinois, Chicago Campus, Chicago, Illinois
November 1983

100 Years of Architectural Drawings in Chicago
Illinois Bell Telephone Company, Chicago, Illinois
November 1983

Tops and Members' Work
The Chicago Architectural Club
The Art Institute of Chicago, Chicago, Illinois
November 1983

The Art of Design
University of Milwaukee Fine Arts Galleries, Milwaukee, Wisconsin
October 1983

The Chicago Architectural Club
The Art Institute of Chicago, Chicago, Illinois
August 1982

Chicago Architects Design
The Art Institute of Chicago, Chicago, Illinois
October 1982

Helmet Jahn
School of Architecture, Yale University, New Haven, Connecticut
November 1982

Contemporary Chicago Architecture
Festival of the Arts
Northern Illinois University, DeKalb, Illinois
November 1982

Helmut Jahn
The Fort Worth Art Museum, Fort Worth, Texas
1981

Chicago Architectural Drawing
Frumkin & Struve, Chicago, Illinois
January 1981

The Chicago Architectural Club
Graham Foundation, Chicago, Illinois
March 1981

Chicago International Art Exhibit
Navy Pier, Chicago, Illinois
May 1981

Architecture as Synthesis
Harvard University Graduate School of Design, Cambridge, Massachusetts
December 1981

Arnold W. Brunner Memorial Prize and Exhibit
American Academy and Institute of Arts and Letters, New York, New York
1981/82

City Segments
Walker Art Center, Minneapolis, Minnesota
April 1980

Late Entries to the Chicago Tribune Competition
Museum of Contemporary Art, Chicago, Illinois
May 1980

The Presence of the Past
Biennale, Venice, Italy
June 1980
Paris, France; San Francisco, California
1981–1982

The Architectural Process
Young-Hoffman Gallery, Chicago, Illinois
August 1980

Designing Today for Tomorrow
Architectural League of New York, New York
Participant in symposium
October 1980

25th Annual Progressive Architecture Awards Jury
New York, New York
Participant
1979

Townhouses
Walker Art Center, Minneapolis, Minnesota
Participant in group show
December 1978

State of the Art of Architecture/77
Graham Foundation, Chicago, Illinois
Participant in symposium
October 1977

Chicago 7
Member
1977

Exquisite Corpse
Walter Kelly Gallery, Chicago, Illinois
Participant in group show
December 1977

AIA Chicago Chapter Awards
The Art Institute of Chicago
1974 – 1986

Bibliography

参考文献

Books

100 Contemporary Architects and their Drawings. Abrams, 1991.

Allen, Gerald & Oliver, Richard. *Architectural Drawing: The Art and the Process.* New York: Whitney Library of Design (imprint of Watson-Guptill Publications) and London: The Architectural Press Ltd, 1981.

American High-Rise Buildings. Tokyo: A + U Publishing Co. Ltd, 1988.

American Skyscrapers/Amerikkalaisia Pivenpiirtaejiae. Exhibition catalog. Finland, 1992.

Architecture Chicago Vol. 6: The Divine Detail. Chicago: Chicago Chapter, American Institute of Architects, 1988.

Arnell, Peter & Bickford, Ted (eds). *A Tower for Louisville: The Humana Competition.* New York: Rizzoli International Publications Inc., 1982.

Arnell, Peter & Bickford, Ted (eds). *Southwest Center: The Houston Competition.* New York: Rizzoli International Publications Inc. 1983.

Blaser, Werner. *After Mies.* Von Reinhold Co., 1977.

Blaser, Werner. *Elemental Building Forms.* Germany: Beton-Verlag, 1982.

Blaser, Werner. *Filigree Architecture.* Germany: Wepf & Co., 1980.

Blaser, Werner. *Mies van der Rohe: Continuing the Chicago School of Architecture.* Germany: Birkhauser Verlag, 1981.

Casari, Maurizio & Pavan, Vincenzo (eds). *New Chicago Architecture.* New York: Rizzoli International Publications Inc., 1981.

Chicago and New York: Architectural Interactions. The Art Institute of Chicago, 1984.

Chicago Architects Design: A Century of Architectural Drawings for the Art Institute of Chicago. The Art Institute of Chicago/New York: Rizzoli International Publications Inc., 1982.

Chicago Architecture. Chicago: Chicago Chapter, American Institute of Architects, 1991.

The Chicago Architectural Journal vol. 1. Chicago: The Chicago Architectural Club/New York: Rizzoli International Publications Inc., 1981.

The Chicago Architectural Journal vol. 2. Chicago: The Chicago Architectural Club/New York: Rizzoli International Publications Inc., 1982.

The Chicago Architectural Journal vol. 3. Chicago: The Chicago Architectural Club/New York: Rizzoli International Publications Inc., 1983.

The Chicago Architectural Journal vol. 4. Chicago: The Chicago Architectural Club/New York: Rizzoli International Publications Inc., 1984.

The Chicago Architectural Journal vol. 5. Chicago: The Chicago Architectural Club/New York: Rizzoli International Publications Inc., 1985.

The Chicago Architectural Journal vol. 6. Chicago: The Chicago Architectural Club/New York: Rizzoli International Publications Inc., 1987.

City-Projekte. Berlin, 1993.

Competitions vol. 1. Louisville, Kentucky: 1991.

Contemporary American Architects. Germany: Benedikt Taschen GmbH, 1993.

Contemporary Masterworks. St James Press, 1992.

De Vito, Alfredo. *Innovative Management Techniques for Architectural Design and Construction.* New York: Whitney Library of Design/Watson-Guptill Publications, 1984.

Diamonstein, Barbara Lee (ed.). *American Architecture Now II.* Forward by Paul Goldberger. New York: Rizzoli International Publications Inc., 1985.

DLW Nachrichten 65/1985. Germany: DLW Architektenforum, 1985.

Emanuel, Muriel (ed.). *Contemporary Architects.* St Martin's Press, 1980.

Fraser, Iain & Henmi, Rod. *Envisioning Architecture—An Analysis of Drawing.* 1994.

From Mars to Main Street: America Designs. Exhibition catalog. Washington, DC: 1993.

Für Sport und Mode: Munich Order Center. Germany: Institut fur Internationale Architektur-Dokumentation, 1993.

Glibota, Ante (ed.). *10 Years of Chicago Architecture.* Paris: Paris Art Center, 1985.

Glibota, Ante (ed.). *Helmut Jahn Modern Romantic.* Paris: Paris Art Center, 1987.

Goldberger, Paul. *The Skyscraper.* New York: Alfred A. Knopf, 1981.

Grube, Pran & Schultze. *100 Years of Architecture in Chicago.* Follet Co., 1976.

Helmut Jahn 1982–1992. Tokyo: A + U Publishing Co., 1992.

Helmut Jahn Airports. Basel: Birkhäuser Verlag, 1991.

Helmut Jahn—Extra Edition. Tokyo: A + U Publishing Co. Ltd, 1986.

Herbert, Daniel M. *Architectural Study Drawings.* 1993.

Hotel Kempinski—Extra Edition. Tokyo: A + U Publishing Co. Ltd, 1995.

Huxtable, Ada Louise. *The Tall Building Artistically Reconsidered: The Search for a Skyscraper Style.* New York: Pantheon Books, 1984.

Ideas and Ideologies in the Late Twentieth Century. American Architecture, Paul Heyer, 1993.

International Architecture Yearbook. Melbourne, Australia: The Images Publishing Group, 1995.

Jahn, Helmut, Kraus, Patrick & Jeschenko, Rudolf. *Helmut Jahn, Building Identity.* Düsseldorf, Germany: ECON, Executive Verlags GmbH, 1992.

Jencks, Carl. *Late Modern Architecture.* New York: Rizzoli International Publications Inc., 1980.

Bibliography Continued

Joedicke, Joachim Andreas (ed.). *Helmut Jahn—Design of a New Architecture.* Stuttgart, Zurich: Karl Kramer Verlag, 1986.

Klotz, Heinrich (ed.). *Die Revision der Moderne: Postmoderne Architektur 1960–1980.* Germany: Prestel, 1984.

Klotz, Heinrich (ed.). *Vision der Moderne—Das Prinzip Konstruktion.* Frankfurt: Deutsches Architekturmuseum/Prestel-Verlag, 1986.

Klotz, Heinrich. *Jahrbuch Fur Architektur 1981/82.* Germany: Deutsches Architektur Museum/Friedr. Viewg. & Sohn, 1981.

Klotz, Heinrich. *Jahrbuch Fur Architektur 1983.* Germany: Deutsches Architektur Museum/Friedr. Viewg. & Sohn, 1983.

Klotz, Heinrich. *Moderne und Postmoderne.* Germany: Friedr. Viewg. & Sohn, 1984.

Krantz, Les (ed.). *American Architects.* New York, 1989.

Laine, Christian K. (ed.). *The Chicago Architectural Annual 1985.* Chicago: Metropolitan Press Publications, 1985.

Lampugnani, Vittorio Magnago. *Architecture in Our Century in Drawings: Utopia and Reality.* Germany: Verlag Gerd Hatje, 1982.

Lauber, M. (ed.). "Beyond the Modern Movement." *Harvard Review,* vol. 1. MIT Press, 1980.

Mardaga, Pierre (ed.). *Biennale de Paris Architecture 1985.* Paris, France.

Miller, Nory. *Helmut Jahn.* New York: Rizzoli International Publications Inc., 1986.

Münchener Order Center MOC: Architekten Murphy/Jahn. Munich: Stahl Informations Zentrum, 1993.

Nelson, Susan R. (ed.). *Groundworks: North American Underground Projects 1980 to 1989.* Washington, DC: 1989.

New York Architektur. Munich: Prestel Verlag, 1989.

Portoghesi, Paolo. *Postmodern: The Architecture of the Postindustrial Society.* New York: Rizzoli International Publications Inc., 1984.

Quaternario 90. Milan, Italy: Electa, 1990.

The Sky's The Limit. Chicago: Art Institute of Chicago, 1991.

A Style for the Year 2001. Tokyo: Shinkenchiku Co. Ltd, 1985.

Sourcebook of Contemporary North American Architecture. New York: Van Nostrand Reinhold, 1989.

Thorndike, Joseph J. Jr (ed.). *Three Centuries of Notable American Architects.* New York: American Heritage Publishing Co. Inc., 1981.

Vision of the Modern. London: UIA Journal of Architectural Theory and Criticism, 1988.

Wirth, Harry J. *The Art of Design 2.* Milwaukee, Wisconsin: 1993.

Periodicals

"1990 AISC Architectural Awards of Excellence." *Modern Steel Construction* (November/December 1990). (United Airlines, One Liberty Place)

"2000 and Beyond." *Progressive Architecture* (December 1985). (State of Illinois Center)

"27th Annual PA Awards." *Progressive Architecture* (January 1980).

"750 Lexington Avenue at 59th Street." *Real Estate Forum* (New York, December 1986).

"88 Field Street." *Architect & Builder* (Johannesburg, July 1986). (362 West Street)

"A Pioneer of Chicago's West Loop." *Building Design and Construction* (September 1988). (Northwestern Atrium Center)

"A Skyscraper on the Corner." *l'Arca* (June 1993). (Fountain Square West)

"A Tale of Four New Towers and What They Tell of Trends." *Architecture* (May 1988). (One Liberty Place)

"A Tall Order—The 31-Story Oakbrook Terrace Tower." *DuPage Profile* (January 21/22, 1987). (Oakbrook Terrace Tower)

"A Third Philadelphia Plan—A Critical View." *Architecture* (October 1988). (Liberty Place)

"Abu Dhabi Conference City." *Progressive Architecture* (January 1977).

"Abu Dhabi/UAE." *Domus* (Milan, June 1979).

"Aerger mit der Presse." *Newmag* (Germany, February 1992).

"Amerika im Streiflicht." *Schweizer Ingenieur und Architekt* (Switzerland, no. 1/2, January 11, 1990). (United Airlines Terminal)

"Analysis of Planned and Completed Projects by C.F. Murphy Associates." *Bauen + Wohnen* (Zurich, September 1974).

"And Now the Tallest of the Tall." *Time* (December 2, 1985). (Television City)

"Arch Angles." *Architecture Journal* (vol. 183, no. 8, February 1986). (Galveston Arches)

"Architects for the 1990's." *Fortune* (June 22, 1987). (North Western Terminal, Park Avenue Tower)

"Architecture as a Corporate Asset." *Business Week* (October 4, 1982). (Addition to the Chicago Board of Trade)

"Architecture Peak Performance." *Chicago Tribune* (October 20, 1991).

"Architecture: Chicago's Skyscraping Romantic—Helmut Jahn." *Interview* (September 1983). (North Western Terminal, Bank of the Southwest Tower)

"Architecture: Helmut Jahn." *Architectural Digest* (March 1984). (Eagle River Wisconsin House)

"Architecture—The Temple of Marketing." *The New Republic* (October 26, 1987). (United Airlines Terminal)

"Architektur: Die grossen Wurfe von morgen" *Stern* (Germany, November 1989).

"Argonne Program Support Facility." *Progressive Architecture* (April 1979).

"Auraria Learning Center." *AIA Journal* (September 1979).

"Auraria Library." *A + U* (Tokyo, October 1978).

"Auraria Library." *Architectural Record* (November 1977).

"Auraria Library." *Domus* (Milan, July 1977).

"Auraria Library." *Inland Architect* (August 1976, November 1976).

"A Babel of New Towers." *Chicago* magazine (April 1984).

"Barone von High Tech." *Epoca* (November 1987). (United Airlines Terminal, State of Illinois Center, O'Hare Rapid Transit Station)

"The Battle of Starship Chicago." *Time* (Design section, February 4, 1985). (State of Illinois Center)

"A Berlin Story." *Interiors* (May 1993).

"The Best Building in Town?" *San Diego* magazine (October 1989). (Great American Plaza)

"The Best of '87—Architecture." *Business Week* (January 11, 1988). (One Liberty Place, United Airlines Terminal)

"Best of Chicago—Best New Skyscraper." *Chicago* magazine (January 1983). (One South Wacker)

"Big Jahn, Offices, Chicago." *Architectural Review* (London, April 1981). (One South Wacker, Xerox Centre)

"Board of Trade Addition: A New Synthesis." *Inland Architect* (September 1980).

"A Bold New Breed of Buidings is Reaching Towards the Skies." *Smithsonian* (October 1983). (One South Wacker, Xerox Centre, Bank of the Southwest Tower)

"The Building of the Year 2000." *Inland Architect* (May 1980).

"Building on a Tradition of Excellence." *Advertising Age* (March 1983).

"Building the New Germany." *Blueprint* (April 1993).

"C.F. Murphy Associates Five Projects." *GA Document 1* (Tokyo, Spring 1980).

"Can Chicago Architecture Live Up to Its Past?." *The New Art Examiner* (January 1984). (One South Wacker, State of Illinois Center, Addition to the Chicago Board of Trade, Xerox Centre, North Western Terminal)

"CFMA Office Building." *A + U* (Tokyo, April 1980).

"CFMA Portfolio." *Domus* (Milan, April 1978).

"CFMA Portfolio." *L'Architectura* (Milan, December 1977).

"CFMA Portfolio." *Progressive Architecture* (July 1978).

"CFMA Portfolio." *The Architectural Review* (London, October 1977).

"The Changing Skins of Helmut Jahn." *AIA Journal* (October 1981).

"Chicago Architecture After Mies." *Critical Inquiry* (University of Chicago, Winter 1979).

"The Chicago Chapter/AIA 1980 Distinguished Building Awards." *Inland Architect* (December 1980).

"Chicago Design: In Search of a New Order." *Interiors* (May 1983). (Murphy/Jahn)

"Chicago Library Entries Unveiled." *Progressive Architecture* (July 1988). (Chicago Public Library Competition)

"Chicago Skyscrapers: Romanticism and Reintegration." *Domus* (September 1984). (Addition to the Chicago Board of Trade)

"Chicago Story..." (Tribune Competition) *Horizon* (July 1980).

"Chicago Team Wins Minnesota Competition" *Progressive Architecture* (April 1977).

"Chicago". *Merian* (Germany, August 1990).

"Chicago-Design." *Town & Country* (September 1990).

"Chicago." *Departures* (October/November 1991).

"Chicago." *Holland Herald* (The Netherlands, vol. 17, no. 12, 1982). (Xerox Centre, State of Illinois Center, North Western Terminal)

"Chicago." *Progressive Architecture* (June 1980).

"Chicago: Architecture City." *Habitat Ufficio* (Milan, April/May 1993). (Murphy/Jahn, Munich Airport Center)

"Chicago: La Griglia L'Infinito." *Domus* (March 1982). (Addition to the Chicago Board of Trade)

"Chicago—Home of Many Greats." *CCAIA Focus* (October 1991).

"City Segments/Exhibition." *Inland Architect* (June 1980).

"Color and Geometry in Wisconsin Woods." *Architecture* (May 1984).

"Competitors Helmut Jalm." *Competitions* (vol. 4, Spring 1994).

"The Constructional Ideas of Helmut Jahn: Paris Art Center Exhibit." *Inland Architect* (January/February 1988).

"Crosby Kemper Arena." *AIA Journal* (April 1976).

"Crossroads Berlin." *The Architectural Review* (January 1993).

Bibliography Continued

"De architectuur van het succes." *Archis* (Amsterdam, May 1986). (State of Illinois Center, Addition to the Chicago Board, One South Wacker, Park Avenue Tower, Humana Competition)

"A Decade for Architecture." *Dialogue* (March 1992).

"Deja Vu: Helmut Jahn in Cincinnati." *Inland Architect* (September/October 1990).

"Der Grosse Hochmut—Helmut Jahn" *Bunte Heft 1* (Germany, December 29, 1988).

"Der Messeturm in Frankfurt." *DBZ* (Germany, January 1992).

"Der Messeturm." *Pflasterstrand* (Germany, September 1990).

"Der Schnellste Colt Imganzen Westen: Jahn-Bau State of Illinois Center in Chicago." *Der Spiegel* (Germany, no. 24, June 10, 1985). (State of Illinois Center)

"Der Turm zeugt vom asthetischen Willen der Stadt." *Galleria* (February 1989). (Messe Frankfurt)

"Der Wolkenkratzer." *German Esquire* (July 1988). (Messe Frankfurt, State of Illinois Center, Xerox Centre, O'Hare Rapid Transit Station, One South Wacker)

"Design Direction: Looking for What is Missing." *AIA Journal* (May 1978).

"Design/Projects Feature: To Be Continued." *Progressive Architecture* (March 1990). (750 Lexington, Park Avenue Tower, One Liberty Place, Messeturm/Messehalle, 120 N. LaSalle, Ku-Damm 70, North Loop, Raffles Place and Collier Quay, 1111 Brickell, Victoria City, Munich Airport Center, JFK Terminal/American Airlines)

"Designs for Living." *Newsweek* (November 1978).

"Design—The Best of 1988: United Airlines Terminal O'Hare International Airport, Helmut Jahn Architect." *Esquire* (December 1988).

"Die Grossen Fun." *Scala* (Italy, September/October 1992).

"Don't talk build." *Living* (February 1993). (Sony Center Berlin)

"Donald Trump's Lofty Ambition." *Newsweek* (December 2, 1985). (Television City)

"Due progetti di Murphy/Jahn." *l'Arca* (January 1993). (Yokohama Waterfront PJ, Tokyu-Shibuya Tower)

"Eauen Sie mir ein Buerohaus in dem Menschen gern arbeiten!" *P.M.* magazine (Germany, January 19, 1990). (State of Illinois Center)

"Echos of the Past Visions for the Present—Best of '87." *Time* (January 4, 1988). (United Airlines Terminal)

"The Economics of Image Building." *Building Design and Construction* (March 1989). (Liberty Place, Oak Brook Terrace Tower)

"Ein Amerikaner am Rhein: Mannheimer Versicherungsbau von Helmut Jahn." *STEINtime* (Germany, September 1992).

"Ein Mann Baut Bunt." ("A Man Builds Bountifully.") *Stern* magazine (Germany, no. 40, September 26, 1985). (Southwest Center, State of Illinois Center, One South Wacker, Shand Morahan Plaza, O'Hare Rapid Transit Station, Oakbrook Post Office, Penn Yards)

"Eingangsbau und Messehalle 1 in Frankfurt/Main" *Glasforum* (Germany, December 1989).

"El encanto de la imitacion." *Arquitectura Viva* (Buenos Aires, March 1994). (Liberty Place)

"Emerging Skyline." *Domus* (Milan, February 1986). (Television City, South Ferry Plaza)

"European Forecast." *Architecture* (September 1990).

"The Excitement Is Building." *Express* (June 1981).

"Facts About: The Terminal for Tomorrow." *Vis a Vis* (August 1987). (United Airlines Terminal)

"Flughafen O'Hare in Chicago —Terminal 1." *Architektur* (Germany, July 1991).

"The Flash Gordon of American Architecture." *Architecture Bulletin* (June 1983).

"Form Follows Fantasy" *Time* (Design section, February 17, 1986).

"Fourth District Courts Building/John Marshall Courts Building." *A + U* (September 1979).

"Frankfurt Campanile." *Architectural Record* (February 1993).

"Frankfurt's Messeturm." *l'Arca* (September 1990).

"Genesis of a Tower: Helmut Jahn Drawings." *AIA Technology* (Fall 1983). (Bank of the Southwest Tower)

"Giganten Desunendlichen." *Ambiente* (Germany, January 1992).

"Glass Highlights Design of Bank/Hotel Complex." *Building Design and Construction* (March 1983). (First Source Center)

"The Glittering City of Helmut Jahn." *Chicago* magazine (February 1986).

"Going for the Glitz." *House and Garden* (March 1988). (United Airlines Terminal, One Liberty Place)

"A Grand Gateway." *Progressive Architecture* (November 1987). (United Airlines Terminal)

"H. Roe Bartle Exhibition Hall." *Detail* (Germany, July/August 1980).

"Half Time—Cerebrating 75 Years of Chicago Architecture." Exhibition catalog. Arts Club of Chicago (January 1992).

"Hauptsache Kultur-Berlin: Verkauf Der Hauptstadt." *Forbes* (April 1992).

"Helmut Jahn (Murphy/Jahn)." *A + U* (Tokyo, November 1983). (Addition to the Chicago Board of Trade, One South Wacker, State of Illinois Center, 701 Avenue South, 362 West Street, Bank of the Southwest Tower, 11 Diagonal Street, MGIC Plaza East, John Deere Harvester Works Office, Shand Morahan Office Building, Greyhound Terminal, Chicago-O'Hare International Airport Development Program, North Western Terminal, Wilshire/Midvale)

"Helmut Jahn El Liberty Place de Filadelfia." *Diseno y Decoracion en la Argentina*. (June 1993).

"Helmut Jahn Four Towers 1978–82." *Architectural Review* (no. 7/8, London, 1983). (Addition to the Chicago Board of Trade, One South Wacker, Humana, Bank of the Southwest Tower)

"Helmut Jahn in Mannheim." *Office Design* (April 1993).

"Helmut Jahn Krafzt Wolken." *Frankfurter Allgemeine* magazine (Germany, December 12, 1986). (Messe Frankfurt, State of Illinois Center, One South Wacker, Shand Morahan Plaza, 70 Lexington, O'Hare Rapid Transit Station)

"Helmut Jahn mit Wolken-Kratzen zum Erflog." *Das Beste aus Reader's Digest* (December 12, 1986). (Messe Frankfurt)

"Helmut Jahn on Architecture as Synthesis." *Harvard Graduate School of Design News* (Winter 1982).

"Helmut Jahn Puts the State Under Glass." *Chicago* magazine (July 1980).

"Helmut Jahn Romances New York A Tough Town." *Crain's Chicago Business* (August 1, 1988). (City Spire, 750 Lexington, 425 Lexington, Park Avenue Tower, 300 E. 82nd St)

"Helmut Jahn State of Illinois Center in Chicago." *Architektur* (Germany, October 1991).

"Helmut Jahn Topples the Box." *Architect & Builder* (Johannesburg, June 1982). (Addition to the Chicago Board of Trade, North Western Terminal, No. 11 Diagonal Street, St Mary's Athletic Facility, Kemper Arena, Xerox Centre, State of Illinois Center)

"Helmut Jahn's Latest Glass-Clad Building in Chicago." *Architect & Builder* (Johannesburg, November 1984). (One South Wacker)

"Helmut Jahn's Tall Buildings." *Art* (October 1990).

"Helmut Jahn, C.F. Murphy Associates." *GA Document 1* (April 1987). (State of Illinois Center, Xerox Centre, Chicago Board of Trade Addition, One South Wacker, North Western Terminal)

"Helmut Jahn." *Bauwelt* (Germany, May 1992).

"Helmut Jahn." *GA Document GA International '89* (1989). (Messeturm)

"Helmut Jahn." *Gentlemen's Quarterly* (September 1983). (One South Wacker, North Western Terminal)

"Helmut Jahn." *Mensch & Buro* (Germany, March 1990).

"Helmut Jahn: Building for the Future." *The Rotarian* (November 1992).

"Helmut Jahn: The Building of a Legend." *New Art Examiner* (November 1987).

"Helmut Jahn—A La Conquete D'Un Monde Nouveau." *Connaissance des Arts* (Paris, December 1987). (North Western Terminal, United Airlines Terminal, Messe Frankfurt)

"Helmut Jahn—Messeturm." *Bunte* (Germany, March 1994).

"High Flight—United Gambles and Wins at O'Hare." *Inland Architect* (September/October 1988). (United Airlines Terminal)

"High Rise Projects." *Bauen + Wohnen* (June 1981). (Xerox Centre, De La Garza Career Center, Chicago Board of Trade Addition, One South Wacker, North Western Terminal, State of Illinois Center)

"High-Rise Hard Sell." *New York* magazine (March 11, 1985). (125 Lexington, Park Avenue Tower)

"High-Tech Expansion: Chicago-O'Hare International Airport Development Program." *Architectural Record* (May 1985).

"High-Tech: Britain v. America." *The Architectural Review* (August 1983). (Argonne Program Support Facility, First Source Center)

"The Hitachi Tower and Caltex House." *l'Arca* (March 1994).

"Hochhaus Messe Frankfurt." Bibliographisches Institut & F.A. Brockhaus AG Band 10, 1989.

"Hochhauser fur Frankfurt." *Baumeister* (January 1987). (Messe Frankfurt)

"Hot Frankfurters" *Commerce in Germany* (Germany, no. 6, September 1989). (Messe Frankfurt)

"Hotel Hyatt Regency Roissy." *Techniques & Architecture* (Special edition, 1993).

"How Helmut Jahn is Changing the Face of Chicago." *Crain's Chicago Business* (November 9, 1981).

"I want to be in America." *German Vogue* (Germany, April 1992).

"Illinois Center." *Architectural Review* (December 1980).

"Im Bauch der Riesen." *Architektur Innenarchitektur Technischer Ausbau* (Germany, October 1989).

"Inside Helmut Jahn's State of Illinois Triumph—A Search for Excellence." *Interiors* (Cover story, May 1985). (State of Illinois Center, CTA Station—O'Hare, United Airlines Terminal, Murphy/Jahn Cupola, Plaza East Office Center, 425 Lexington, City Center, Park Avenue Tower)

"Inside the Livable City." *Inland Architect* January/February 1989). (Addition to the Chicago Board of Trade, State of Illinois Center)

"An Island Outside the Paris Gates" *l'Arca* (July/August 1991). (Hyatt Regency Roissy, Ku-Damm 70)

Bibliography Continued

"It's What's Outside That Counts" *Inland Architect* July/August 1991). (Wilshire/Westwood)

"Jahn Building Identities." *High Touch* (no. 2, 1990). (State of Illinois Center)

"Jahn in the Loop." *On the Boards Architecture* (February 1990). (550 Madison, Savings of America Tower, Wells & Lake Office Tower, North Loop Block 37, Project J)

"Jahn's Bold Designs Taking Root in West Suburban Soil." *Crain's Chicago Business* (March 24, 1986). (MetroWest, Oakbrook Terrace Tower)

"Jahn's Chicago 1: Board of Trade", "Jahn's Chicago 2: One South Wacker." The Sixth Annual Review of New American Architecture. *AIA Journal* (May 1983).

"Jahn's Designs Shatter the Box." *Engineering News Record* (October 8, 1981).

"Jeu d'Enveloppes, Un Rond dans L'eau." *Techniques & Architecture* (Paris, September 1984). (First Source Center, Argonne Program Support Facilities)

"Kansas City Convention Center." *Domus* (Milan, February 1979).

"Kemper Arena." *Architectural Record* (March 1976).

"Kemper Arena." *Detail* (Germany, May/June 1979).

"Kemper Arena." *Domus* (Milan, July 1976).

"Kemper Arena." *Inland Architect* (June 1976).

"Kemper Arena/H. Roe Bartle." *AIA Journal* (March 1979).

"Kempinski Hotel." *l'Arca* (January 1995).

"Ku Damm 70 Berlin" *Progressive Architecture* (April 1989).

"Ku Damm 70, Ku Damm 119, Victoria Berlin and Sony Center Berlin." *Nikkei Architecture* (Tokyo, December 20, 1993).

"La Lumiere Gymnasium." *Domus* (January 1980).

"La Torre Della Fiera Di Francoforte" *Abitare* (Milan, no. 274, May 1989). (Messe Frankfurt)

"Late Entries." *Progressive Architecture* (June 1980).

"Late Postmodern—The End of Style." *Art in America* (June 1987).

"Latest Murphy/Jahn." *World Architecture* (London, March 1994).

"Look What Landed in the Loop." *Architecture* (November 1985). (State of Illinois Center)

"Made by Helmut Jahn." *Architektur Innenarchitektur Technischer Ausbau* (Germany, December 1992).

"Man of Dreams." *BMW magazine* (Germany, January 1993).

"The Master Builder." *GQ* (Cover story, May 1985). (State of Illinois Center, Xerox Centre, One South Wacker)

"Messehalle 1 Frankfurt/Main" *Dokumentation Detail* (Germany, August/September 1989).

"Messeturm and Messe Halle." *Nikkei Architecture* (Tokyo, October 1991).

"The Messe Tower/Messe Halle in Frankfurt." *l'Industria Delle Costruzioni* (March 1994).

"Messeturm Frankfurt." *Ambiente* (Germany, October 1992).

"Messeturm." *Architektur Innenarchitektur Technischer Ausbau* (Germany, cover story, April 1989).

"Metaalnijverheid met architect Helmut Jahn in Rotterdam." *UGM* (May 1991).

"Michigan City Library." *Building Design and Construction* (December 1977).

"Michigan City Library; St Mary's Athletic Facility." *GA Document* (Special Issue 1979–1980).

"Minnesota II." *Architecture Minnesota* (January/February 1978).

"Minnesota II." *Engineering News Record* (March 1977).

"Minnesota II." *Progressive Architecture* (January 1979).

"MOC—Munchner—Order Center." *DBZ* (Germany, February 1994).

"Monroe Centre." *Inland Architect* (October 1977).

"Monroe Centre." *Progressive Architecture* (January 1978).

"Movement and Color as Themes." *Architecture* (May 1987). (O'Hare Rapid Transit Station)

"Munich Order Center and Mannheimer Lebensversicherung." *Aprire* (no. 2, 1993).

"Munich Order Center." *l'Arca* (February 1994).

"Murphy y Jahn en la feria." *Arquitectura Viva* (Buenos Aires, no. 8, October 1989). (Messeturm)

"Murphy/Jahn Profile." *Inland Architect* (May/June 1992).

"Murphy/Jahn." *A + U* (Tokyo, February 1983). (First Source Center, Argonne National Laboratories Chicago Branch/Department of Energy, Area 2 Police Center, United States Post Office)

"Murphy/Jahn." *A + U* (Tokyo, January 1986). (State of Illinois Center, 10 Columbus Circle)

"Murphy/Jahn." *World Construction and Engineering* (vol. 2, no. 10, Tokyo, June 1983). (Bank of the Southwest Tower, Addition to the Chicago Board of Trade)

"Murphy/Jahn." *World Construction and Engineering* (vol. 2, no. 12, Tokyo, August 1983). (11 Diagonal St, One South Wacker)

"Murphy/Jahn: State of Illinois Center." *A + U* (January 1984).

"Murphy/Jahn—Recent works and Projects." *GA Document 7* (September 1983). (State of Illinois Center, Oak Brook Post Office, First Source Center, Argonne Program Support Facility, Addition to the Chicago Board of Trade, One South Wacker, Bank of the Southwest Tower, 11 Diagonal Street, 701 Fourth Avenue South, Greyhound Terminal, 362 West Street, Americana Plaza, Shand Morahan Plaza, Wilshire/Midvale, North Loop Block F)

"New Chicago Architecture." *Inland Architect* (May/June 1983). (Bank of the Southwest Tower)

"New Chicago Projects" *Metropolitan Review* (September/October 1989). (North Loop Block 37, Lake and Wells Office Tower, Wells and Lake Office Tower)

"New Departures." *Designers' Journal* (vol. 42, November 1988). (United Airlines Terminal)

"New Designs and Directions at C.F. Murphy Associates." *Architectural Record* (July 1979).

"A New Era Dawns for First Source." *Corporate Design* (November/December 1982). (First Source Center)

"New Wall System Weds Diverse Materials." *Building Design and Construction* (March 1986). (Shand Morahan Plaza)

"A New Trump Card." *Inland Architect* (January 2, 1986). (Television City)

"North Loop Block 37" *Inland Architect* (September/October 1989).

"North Loop Block 37." *l'Arca* (May 1991).

"North Western Center." *Arquitectura* (Buenos Aires, February 1993).

"North Western Terminal." *Bunte* (Germany, no. 6, March 2, 1983).

"Nur Mut zum Streit schafft die ganz grosse Architektur." *Inpulse* (Germany, March 1992).

"The O'Hare Station—Last Link to the Loop." *Inland Architect* (July/August 1985).

"O'Hare United Terminal" *Bauwelt* (Germany, no. 12, March 24, 1989).

"Oben Hui Unten Pfui." *Bauwelt* (Germany, July 1993).

"On Visability Travel and Transformation The Questions of Northwestern Atrium." *Inland Architect* (May/June 1988).

"One Liberty Place, Two Liberty Place, City Spire, Messe Tower." *Real Estate Forum* (May 1988).

"One South Wacker." *Inland Architect* (September 1980).

"Opinion." *Inland Architect* (November 1974).

"Orizzontale/Verticale." *l'Arca* (September 1991). (1111 Brickel Avenue, Project J, Stuttgart Berliner Platz)

"Pa Vilhao de Exposicoes em Munique." *Arquitectura Urbanisma* (Buenos Aires, October/November 1993). (Munich Order Center)

"Pahlavi National Library." *Domus* (Milan, August 1978).

"Prima Donnas on the Landscape—Paul Goldberger on the Indulgence of Postmodernism." *The New Art Examiner* (February 1989). (One Liberty Place)

"Private Residence Eagle River Wisconsin." *GA Houses 17* (January 1985).

"Profile: Chicago Seven, The Exquisite Corpse." *A + U* (Tokyo, July 1978). (Helmut Jahn, C.F. Murphy Associates)

"Prominent Persons in High Tech Architecture—Helmut Jahn." *SD*—*Space Design* (January 1985). (Parktown, Chicago O'Hare International Airport Development Program, 362 West Street)

"Recent Works of Chicago 7." *Architectural Review* (June 1980).

"Reordering the Suburbs." *Progressive Architecture* (May 1989). (Oak Brook Terrace Tower)

"Residence in Eagle River." *A + U* (Tokyo, July 1985).

"Residence—Eagle River Wisconsin U.S.A. Murphy/Jahn." *L'Architecture D'Aujourd'hui* (April 1984).

"Rise to the Top—The Success of Chicago Architect Helmut Jahn." *Metropolis New York* (March 1986).

"Rust-Oleum Corporate Headquarters." *Baumeister* (Germany, September 1980).

"The Savings of America Tower." *l'Arca* (May 1994).

"Screenwall." *Deutsche Bauzeitung* (Germany, July 1991). (Kempinski Hotel)

"Shall We Dance—The Suburban Context of Oakbrook Terrace Tower." *Inland Architect* (November/December 1988).

"The Sky's the Limit." *Newsweek* (November 8, 1982). (Bank of the Southwest Tower, Xerox Centre, Bank of the Southwest Tower, Xerox Centre, State of Illinois Center, North Western Terminal)

"Skyscraper View." *Design Quarterly* (no. 140, Summer 1988). (One Liberty Place)

"Skyscrapers and the City." *Chicago History* Winter 1983/84). (State of Illinois Center, Bank of the Southwest Tower)

"Skyscrapers: Above the Crowd." *National Geographic* (February 1989). (One Liberty Place, Helmut Jahn)

"Snake Skin Skyscrapers." *Domus* (Milan, April 1981). (Xerox Centre, Addition to the Chicago Board of Trade, North Western Terminal)

"Soaring Spaces that Celebrate Travel." *Architecture* (May 1988). (United Airlines Terminal 1)

"Sony-Wettbewerb." *Bauwelt* (Germany, October 1992).

Space Design (February 1983). (State of Illinois Center, Argonne Program Support Facility)

"Special Export, World Comes to Chicago to Draft a Tall One." *Commercial Real Estate Magazine Sun Times* (Spring 1994).

Bibliography Continued

"St Mary's Athletic Facility." *Architectural Record* (April 1979).

"Star on Skyline." *San Diego County Los Angeles Times* (October 1991).

"Starchitects." *Chicago Tribune* magazine (January 18, 1987). (State of Illinois Center, Xerox Centre, One South Wacker, Two Energy Center, One Liberty Place, Messe Frankfurt, New York projects)

"State of Illinois Center." *GA Document 13* (Tokyo, November 1985).

"State of Illinois Center." *Progressive Architecture* (February 1981).

"The State of Space." *Inland Architect* (March/April 1987). (State of Illinois Center, 701 Fourth Avenue South)

"Strutture Metropolitane." *l'Arca* (July/August 1992). (United Airlines Terminal, JFK Terminal)

"Tall Tower for Texas." *Time* (Design section, November 8, 1982). (Bank of the Southwest Tower, Addition to the Chicago Board of Trade, Xerox Centre)

Techniques & Architecture (September 8, 1981). (State of Illinois Center, Xerox Centre, One South Wacker, Chicago Board of Trade, North Western Terminal)

"Television City a New York" *l'Arca* (March 1987).

"Television City." *Architect & Builder* (Johannesburg, May 1986).

"Terminal of the Future." *ID* (January/February 1988). (United Airlines Terminal)

"Teutonisches Wunderkind." *Der Spiegel* (Germany, no. 27, July 5, 1982).

"Three Designs by Murphy/Jahn." *Architectural Record* (December 1981). (North Western Terminal Project, Railway Exchange Building Renovation, O'Hare Rapid Transit Station)

"Three New Designs by C.F. Murphy Associates." *Architectural Record* (August 1980).

"Times Square Project New York, New York, USA Design." *GA Document* (Spring 1987).

"Today's Towers: Reaching for New Heights." *American Arts* (May 1983). (Bank of the Southwest Tower, Humana Corporation Competition)

"Tokyo Teleport Town." *l'Arca* (December 1991).

"Toward Romantic Hi Tech." *Architectural Record* (January 1983). (Area 2 Police Headquarters, Argonne Program Support Facility, First Source Center)

"The Tribune Competition 1922/1980." *Inland Architect* (May 1980).

"Turm-Vater Jahn." *Manager Privat* (Hamburg, May 1986). (Television City)

"U-Bahn Station 'O'Hare' in Chicago Il. USA." *Architektur + Wettbewerbe* (Germany, no. 134, June 1988). (O'Hare Rapid Transit)

"U.S. Architectural Firms Scale New Heights in Pursuit of Project Spawned by Asia's Boom." *The Asian Wall Street Journal Weekly* (March 1994).

"United Airlines Terminal: Best of Category." *ID* (Annual Design Review, July/August 1988).

"United Airlines." *The Canadian Architect* (Canada, June/July 1991).

"United Airlines—O'Hare Framed for Tomorrow." *Modern Steel Construction* (no. 5, September/October 1987). (United Airlines Terminal)

"United's Crystal Palace." *Air & Space Smithsonian* (October/November 1988). (United Airlines Terminal)

"Unschuld verloren." *Der Spiegel* (Germany, no. 14, March 31, 1986). (Messe Frankfurt)

"Urbanes Drama." *Der Spiegel* (Germany, August 1992).

"The Uses of Glass." *Progressive Architecture* (March 1989). (United Airlines Terminal O'Hare International Airport)

"Vision in Glass." *Eckelt* (Germany, December 1993).

"Vom MUC zum MOC." *Bauwelt* (Germany, July 1993).

"What's Next?." *AIA Journal* (May 1980).

"Why Europe's New Skyline is Looking So American." *Business Week* (August 15, 1988). (Messe Frankfurt)

"Wie man mit Beton den Himmel erobert: Der Frankfurter Messeturm." *Frankfurter Allgemeine* magazine (Germany, no. 16, vol April 20, 1990, Heft 529).

"Working Abroad." *Architecture* (September 1993).

"Works of Murphy/Jahn." *A + U* (Tokyo, October 1985). (11 Diagonal Street, 701 Fourth Avenue South, Plaza East, College of Du Page, Shand Morahan Office Headquarters, O'Hare Rapid Transit, University of Illinois Agricultural/Engineering Science Building)

"Xerox Centre." *AIA Journal* (May 1981).

"Xerox Centre." *GA Document 3* (April 1981).

"Xerox Centre." *Progressive Architecture* (December 1980).

Acknowledgments

致谢

The work in this book was made **possible** through the efforts of the **Senior Personnel: Sam Scaccia, Philip Castillo, Brian O'Connor, Rainer Schildknecht, Martin F. Wolf, Gordon Beckman, Steven S. Cook, John S Durbrow, Susan Froelich, Scott Pratt, Keith H. Palmer, Dieter Zabel, Edward P. Wilkas, Mark Joshua Frisch, Sanford E. Gorshow, Stephen Kern, Steven M. Nilles, Thomas M. Chambers, Stephen Cavanaugh and Gregg Leoscher; Consulting Architects: Charles Bostick, Fritz Ludwig, Lothar Pascher, Raimond Schöck and Former Senior Personnel: Daniel Dolan, James Goettsch, Robert Goldberg, James M. Stevenson and Dennis Recek.**

Through all the **Clients** who commissioned this work.

Through the cooperative efforts of **Engineers and Consultants** in particular:

STRUCTURAL DESIGN
Burggraf, Weichinger & Partner GmbH
Martin & Associates Group, Inc.
Ove Arup & Partners
Schlaich, Bergermann & Partner
Sobek und Rieger
Thornton Tomasetti

MECHANICAL AND ELECTRICAL DESIGN
Cosentini Associates
Ebener & Partner
Flack + Kurtz Consulting Engineers
IC Consult
Jaros Baum & Bolles
Ove Arup & Partners

LIGHTING DESIGN
Francis Krahe & Associates
L'Observatoire International Inc.
Sylvan R Shemitz & Associates

LANDSCAPING DESIGN
Peter Walker, William Johnson & Partners

Special thanks to **Keith H. Palmer** for coordinating the preparation of this book.

We would like to thank **The Images Publishing Group, Paul Latham and Alessina Brooks,** for their invitation to participate in this monograph series.

Photography Credits:
Wayne Cable: 162 (2,3)
Glenn Cormier: 209 (3)
S. Couturier, Archipress: 203 (10)
Hans Ege: 182 (26); 187 (47, 49)
H.G. Esch: 79 (17)
Engelhardt/Sellin, Aschau I. CH., Germany: 15 (3–5); 16 (9, 10); 19 (15–19); 136 (2); 137 (3–5); 138 (7, 8); 139 (9); 140 (13); 141 (14–16); 175 (3–5); 176 (6–11); 177 (13, 14); 178 (16); 179 (17–20, 25); 201 (3, 4); 202 (5, 6); 203 (9)
Scott Francis, Esto: 90 (3); 93 (2–5); 94 (1); 115 (3, 4)
Peter Mass, Esto: 114 (1)
Courtesy of Josef Gartner & Co.: 73 (3); 74 (6)
Roland Halbe: 73 (3); 74 (4–6); 75 (7, 8); 76 (10); 77 (12); 78 (13–15); 79 (16–21)
Hedrich-Blessing: 103 (4, 5)
Timothy Hursley, The Arkansas Office: 97 (2); 98 (3); 99 (5); 109 (3–5); 110 (7–9); 111 (10); 214 (2); 215 (5); 216 (7); 217 (8, 10–11); 218 (12, 13, 15); 219 (16)
The Images Publishing Group, Tim Griffith: 17 (11); 19 (20); 203 (11)
Tim Griffith: 57 (2); 58 (3); 59 (4, 5); 60 (7, 8); 61 (9, 11); 62 (13, 14); 64 (17); 65 (19)
Helmut Jahn: Front cover; 40 (2); 41 (3); 132 (1); 133 (4, 5); 134 (6–9); 135 (10); 153 (3, 4); 181 (25); 182 (27–29); 183 (30–33); 184 (34–38); 188 (51, 52); 189 (53, 54)
Paul Kivett: 44 (2); 45 (4–6)
George Lambros: 103 (2, 3)
Nathaniel Lieberman: 115 (5)
John Linden: 15 (6, 7); 18 (13, 14); 180 (22)
John McGrail: 104 (2, 4)
Michael Meyersfeld: 122 (2); 123 (3–5)
Gregory Murphey: 166 (6)
Keith H. Palmer: 36 (3); 37 (4–5, 7–8); 38 (2); 39 (3–5); 43 (2); 48 (3); 54 (3, 4); 143 (1); 148 (2, 3); 199 (7, 8); 192 (2); 193 (3); 194 (7)

Keith H. Palmer/James Steinkamp: 23 (2); 25 (3–5); 26 (6–8); 27 (9–11); 28 (2); 29 (3–6); 30 (2, 3); 31 (4, 5); 33 (3–7); 34 (2); 35 (5); 124 (2); 125 (3); 127 (3–7); 128 (2); 129 (3); 164 (2); 165 (3); 166 (4, 5); 220 (2); 221 (3–5); 222 (1, 3)
Ryan Roulette: 208 (1); 209 (4)
Raimund Schoeck: 79 (16, 20)
Sedlacyek, courtesy of Josef Gartner & Co.: 79 (19)
Mark Segal: 167 (7)
Skyline Studio: 113 (4, 5)
Courtesy of Stahl: 79 (19)
James Steinkamp, Steinkamp/Ballogg Photography: 20 (3); 21 (4); 22 (3); 49 (2, 3); 51 (3–5); 66 (1); 67 (1); 69 (3); 81 (3, 4); 82 (2); 83 (1); 87 (3, 4); 89 (3, 5); 95 (1, 2); 100 (2, 3); 101 (2, 3); 110 (6); 159 (2, 4); 160 (2); 196 (2); 197 (3, 4); 207 (9); 211 (3, 4); 212 (5, 6); 213 (8, 9); 217 (9); 218 (13); 222 (2, 4)
Tod Swlechichowski: 75 (8)
Addison Thompson: 84 (2); 85 (3)
Lawrence S. Williams, Inc.: 105 (3); 107 (6–8)
Jay Wolke: 71 (8)
Fernado Urquijo: 203 (7); 204 (8)

Rendering Credits:
Michael Budilovsky: 50 (2); 21 (5–8); 22 (1, 2); 23 (1); 118 (2); 119 (1); 142 (1, 2); 143 (2); 151 (5); 206 (7); 198 (5, 6)
Martin Wolf: 190 (2); 191 (3–5); 193 (4); 194 (5, 6)

Index 索引

Bold page numbers refer to projects included in Selected and Current Works.

1000 West Sixth Street, Los Angeles, California 237

11 Diagonal Street, Johannesburg, South Africa **122**, 233

1111 Brickell Avenue, Miami, Florida **100**, 237

120 North LaSalle, Chicago, Illinois **68**, 237

21 Century Tower, Shanghai, China **50**, 241

300 East 85th Street, New York, New York 235

362 West Street, Durban, South Africa **112**, 234

425 Lexington Avenue, New York, New York **90**, 235

701 Fourth Avenue South, Minneapolis, Minnesota 233

750 Lexington Avenue, New York, New York **92**, 235

9th and Figueroa, Los Angeles, California 236

Abu Dhabi Conference Center City, Abu Dhabi, United Arab Emirates **42**, 230

Alexanderplatz, Berlin, Germany 241

Alliance Airport Terminal—Administration Building, Fort Worth, Texas 246

ANL/DOE Program Support Facility, Argonne, Illinois **28**, 230

Area 2 Police Headquarters, Chicago, Illinois **30**, 230

Augustinerhof-Nuernberg, Nuernberg, Germany **158**, 244

Auraria Library, Denver, Colorado **40**, 228

Bangkok International Airport, Bangkok, Thailand **192**, 248

Bank of Southwest Tower, Houston, Texas **120**, 236

Barnett Center, Jacksonville, Florida **84**, 237

Caltex House, Singapore **56**, 238

Canadian Broadcast Tower, Toronto, Canada 239

Celebration Center, Orlando, Florida **20**, 232

Charlemagne, Brussels, Belgium **142**, 245

Chicago Board Of Trade Addition, Chicago, Illinois **124**, 233

Citizens & Southern Bank, Atlanta, Georgia 239

Cityspire, New York, New York **94**, 235

COD/Learning Resources Center, Glen Ellyn, Illinois 231

Codex Corporate Headquarters, Canton, Massachusetts 231

Columbus Circle, New York, New York **119**, 236

Columbus Pier, Hamburg, Germany 245

Commonwealth Edison District Headquarters, Bolingbrook, Illinois 229

Communication Center FFM, Frankfurt, Germany 244

Credit Lyonnais Bank Nederland, Rotterdam, The Netherlands 238

De La Garza Career Center, East Chicago, Indiana **32**, 229

Dearborn Station, Dearborn, Michigan **180**, 248

Dockland Square, London, England 239

Endless Towers **54**, 241

Europa-Haus, Frankfurt, Germany **160**, 244

Exposition and Office Complex, Chicago, Illinois 231

FAA O'Hare Control Tower, Chicago, Illinois **82**, 240

First & Broadway Los Angeles Civic Center, Los Angeles, California **163**, 242

First Source Center, South Bend, Indiana **170**, 242

Flughafen Koln/Bonn Terminal One, Cologne, Germany **196**, 248

Fort Canning Tower, Singapore **66**, 240

Fountain Square West, Cincinnati, Ohio **80**, 240

Fourth District Courts Building, Maywood, Illinois 228

H. Roe Bartle Exhibition Hall, Kansas City, Missouri 228

Hague Town Hall and Library, Amsterdam, The Netherlands 242

Harold Washington Library Center Competition, Chicago, Illinois **162**, 243

Hitachi Tower, Singapore **56**, 238

Hotel Kempinski/Munich Airport Center/Walkways, Munich, Germany **174**, 247

Hyatt Regency Roissy, Roissy, Charles De Gaule, France **200**, 247

IHZ Tower, Düsseldorf, Germany **67**, 240

Jakarta Communications Tower **48**, 241

Jakarta Tower Kuningan Centre, Jakarta, Indonesia **49**, 241

JFK Consolidated Terminal, Jamaica, New York **210**, 247

John Marshall Courts Building, Richmond, Virginia 228

Kemper Arena Kansas City, Missouri **44**, 228

King Abdulaziz International Airport Competition, Jeddah, Saudi Arabia **204**, 248

Ku-Damm 119, Berlin, Germany **152**, 243

Ku-Damm 70, Berlin, Germany **132**, 243

Kuala Lumpur City Center 1, Kuala Lumpur, Malaysia **83**, 240

La Lumiere Gymnasium, La Porte, Indiana 230

Landesgirokasse Bank Center, Stuttgart, Germany 244

Livingston Plaza, Brooklyn Heights, New York 242

Lufthansa Corporate Headquarters, Köln, Germany **161**, 244

Mannheimer Lebensversicherung Headquarters, Mannheim, Germany **154**, 243

Merchandise Mart Bridge, Chicago, Illinois 246

Messe Hall 1, Frankfurt, Germany 232

Messe Riem Urban Competition, Munich, Germany 232

Messe Tower/Messe Hall, Frankfurt, Germany **72**, 235

Metro West, Naperville, Illinois **116**, 234

Michigan City Public Library, Michigan City, Indiana **38**, 229

Minnesota Government & History Center, St Paul, Minnesota **43**, 229

Monarch Beach Office Building, Dana Point, California 232

Munich Order Center, Munich, Germany **14**, 232

Navy Pier, Chicago, Illinois **22**, 232

The North Loop Block 37, Chicago, Illinois **101**, 238

Northwestern Atrium Center, Chicago, Illinois **108**, 234

O'Hare Rapid Transit Station, Chicago, Illinois **220**, 246

Oakbrook Post Office, Oakbrook, Illinois 230

Oakbrook Terrace Tower, Oakbrook Terrace, Illinois **102**, 235

One America Plaza 237

One America Plaza/Trolley Station, San Diego, California **208**, 247

One Liberty Place, Philadelphia, Pennsylvania **104**, 234

One South Wacker, Chicago, Illinois **126**, 233

Pacific Basin Tower 236

Palace Prince Khaled, Saudi Arabia 232

Pallas Office Building, Stuttgart, Germany **136**, 243

Park Avenue Tower, New York, New York **114**, 234

Parktown Stands, Johannesburg, South Africa 231

Plaza East, Milwaukee, Wisconsin 233

Principal Mutual Life Insurance Company, Des Moines, Iowa **143**, 245

Project J (1, 2, 3), Chicago, Illinois **95**, 238

Reichstagsgebaude Competition, Berlin, Germany 245

Rust-Oleum Corporation Headquarters, Vernon Hills, Illinois **34**, 229

San Diego Convention Center, San Diego, California **23**, 231

Shand Morahan Corporate Headquarters, Evanston, Illinois **168**, 242

Sony Center Berlin, Berlin, Germany **148**, 245

South Ferry Plaza, New York, New York **118**, 236

St Mary's Athletic Facility, South Bend, Indiana **36**, 229

State Of Illinois Center, Chicago, Illinois **164**, 242

Stralauer Platz 35, Berlin, Germany **144**, 245

Stuttgart/Berliner Platz, Stuttgart, Germany 237

Television City, New York, New York 236

Tokyo Teleport Town, Tokyo, Japan **88**, 239

Tokyu-Shibuya Tower, Tokyo, Japan **55**, 241

Two Liberty Place, Philadelphia, Pennsylvania 237

United Airlines Terminal One Complex, O'Hare Airport, Chicago, Illinois **214**, 246

University of Illinois Agricultural Engineering Science Building, Champaign, Illinois 230

Victoria Düsseldorf Competition, Düsseldorf, Germany 242

Victoria/Victoria City Areal, Berlin, Germany 244

Victoria Versicherungsgesellshaft **150**, 244

Wan-Chai Competition, Hong Kong 239

West Loop Pedway, Chicago, Illinois 246

Wilshire/Westwood, Los Angeles, California **96**, 234

Wisconsin Residence, Eagle River, Wisconsin **24**, 231

World's Tallest Building 239

Xerox Centre, Chicago, Illinois **128**, 233

Yokohama Waterfront MM21 Project, Yokohama, Japan **86**, 240

Every effort has been made to trace the original source of copyright material contained in this book. The publishers would be pleased to hear from copyright holders to rectify any errors or omissions.

The information and illustrations in this publication have been prepared and supplied by Murphy/Jahn. While all reasonable efforts have been made to ensure accuracy, the publishers do not, under any circumstances, accept responsibility for errors, omissions and representations express or implied.